Learning and teaching
in school science

Learning and teaching in school science
PRACTICAL ALTERNATIVES

EDITED BY
Di Bentley and Mike Watts

OPEN UNIVERSITY PRESS
Milton Keynes · Philadelphia

Open University Press
12 Cofferidge Close
Stony Stratford
Milton Keynes MK11 1BY

and

242 Cherry Street
Philadelphia, PA 19106, USA

First published 1989

British Library Cataloguing in Publication Data
Learning and teaching in school science: practical alternatives
1. Schools. Curriculum subjects; Science
Teaching
I. Bentley, Di   II. Watts, Mike
507'.1

ISBN 0-335-09514-3
ISBN 0-335-09513-5 Pbk

Library of Congress Cataloging-in-Publication Data
Learning and teaching in school science: practical alternatives
Includes index.
1. Science—Study and teaching. I. Bentley, Di.
II. Watts, Mike.
Q181.P66 1988   507'.1   88-17986
ISBN 0-335-09514-3
ISBN 0-335-09513-5 (pbk.)

Typeset by Rowland Phototypesetting Limited,
Bury St Edmunds, Suffolk
Printed in Great Britain by the Alden Press Limited, Oxford

We dedicate this book to Sam, Siân and Rhian

CONTENTS

Contributors   xi
Preface   xiii
Acknowledgements   xv

1 Learning to make it your own   1
Mike Watts, Di Bentley and Joseph Hornsby

Introduction   1
Passive and active learning   3
Why not encourage pupils to be passive?   5
Science education policy   7
Changes from within science education   10
Evidence from research in science education   10
The effect of examination changes   13
So what is active learning?   14
The chapters of this book   17
Bibliography   19

2 Practicals and projects   21

Introduction   21
What might practical work in school science be?   22
Case study 1: Project work
   John Heaney   23
Case study 2: Project work and practical assessment
   Mary Doherty   27
Case study 3: Practical activity – whose problem is it?
   Philip Naylor   31
Case study 4: Science process investigations
   Di Bentley   35

Summary   38
Bibliography   41

3   Talking and writing for learning   42

Introduction   42
Case study 5: Discussing physics
   Mike Watts   46
Case study 6: Enhancing listening skills in science
   Harry Moore   51
Case study 7: A day in the life of a gas particle
   Pauline Hoyle   54
Case study 8: We wrote a book
   Steve Whitworth   59
Case study 9: Science with rhyme and reason
   Mike Watts   63
Case study 10: Oral assessment in science
   Brigid Bubel   67
Summary   74
Bibliography   79

4   Problem solving   80

Introduction   80
Case study 11: The desert island problem
   Andy Howlett   84
Case study 12: Problem solving – a way of promoting
   scientific skills and processes
   Beverley A. Cussans   91
Case study 13: Solving problems – relevance and creativity
   in science
   Dave Wallwork   94
Case study 14: Problem solving – fuelling up for the start
   Mick Nott   97
Summary   100
Bibliography   102

5   Encouraging autonomous learning   104

Introduction   104
Case study 15: The ABAL project
   Peter Richardson   106
Case study 16: Teaching physics in an all girls school and
   ways of personalizing their learning
   Mary Doherty   110

Case study 17: Science with a soul – learning from friends
  Di Bentley   112
Case study 18: A primary approach to secondary science
  Brian Taylor   117
Summary   120
Bibliography   121

6 **Games and simulations: aids to understanding science**   123

Introduction   123
Case study 19: Playing games in physics
  Linda Scott   125
Case study 20: The Cherry Pie game
  Margaret Davies   129
Case study 21: Sexually-transmitted diseases – AIDS
  Norma White   133
Case study 22: Computer-assisted learning – a biological
  simulation
  Anita Pride   135
Summary   138
Bibliography   140

7 **Using role play and drama in science**   142

Introduction   142
Case study 23: Drama, science and issues of value
  Jon Nixon   144
Case study 24: Educational drama
  Martin Hollins   148
Case study 25: 'X, Y and Z' – a drama-based approach to the
  introduction of genetics in the secondary school
  Hamish Fyfe   153
Summary   157
Bibliography   161

8 **Media and resource-based learning**   162

Introduction   162
Case study 26: Using science education broadcasting as a
  part of the science curriculum
  Phil Munson   166
Case study 27: Vision and science learning experiences
  Robin Moss   170

Case study 28: The classroom use of industrial resource
    material
    Rod Dicker   173
Summary   177
Bibliography   182

9 Summary and Discussion
Mike Watts and Di Bentley   183

Introduction   183
How can we come to terms with active learning?   187
How can we gauge the success of a particular teaching
    method?   189
What of teaching for equal opportunities?   191
Multi-ethnic and anti-racist school science   192
Where to next?   193
A last piece of advice   195
Bibliography   197

Index   199

CONTRIBUTORS

Di Bentley
Brigid Bubel
Bev Cussans
Margaret Davies
Rod Dicker
Mary Doherty
Hamish Fyfe
John Heaney
Martin Hollins

Joseph Hornsby
Andy Howlett
Pauline Hoyle
Harry Moore
Robin Moss
Phil Munson
Philip Naylor
Jon Nixon
Mick Nott

Anita Pride
Peter Richardson
Linda Scott
Brian Taylor
Dave Wallwork
Mike Watts
Norma White
Steve Whitworth

PREFACE

This book is intended to provide some entrée for readers into a series of different teaching approaches. On the whole they are set within science education, although we believe they are of value to teachers outside science and in different parts of the education system. We see them as approaches which are alternatives to the common diet offered in science lessons, and yet practical and feasible for teachers to use. The book is organized as follows. Our first chapter deals with pressures for change. Our central argument is not concerned with science content – this is mostly catered for in the edicts of the National Curriculum. We argue that it is the approach that matters: science teachers need to adopt new and different approaches to teaching and learning. We try to suggest an answer to the question 'Why should I change the way I teach?' In particular we focus on the notion of active learning – a theme that runs through the remainder of the book. In the chapters that follow, we cluster three, four or five case studies around a series of themes. In these studies, our contributors detail the use of a range of teaching methods.

Each contributor has specialist experience in their field. Their contributions vary in length: some tend towards general arguments whereas others tend towards specific approaches. In one or two cases the authors have included original pupil work. As editors, we introduce and summarize each chapter and draw out practical points from each case. The questions addressed in each end-piece are intended to provide a basis for discussion and debate so that the studies can be used as the focus for work in school-based Inset.

The final chapter is our summary of the approaches and their implications for teaching science for the National Curriculum. We try to show how these approaches can be used in different contexts. We have grouped the references and further reading by chapter under the heading of bibliography.

We intend the book as a useful, practical guide to a variety of strategies and classroom activities: a collection of experiences and ideas about different teaching methods which will benefit both trainee and practising teacher. They are designed to appeal to those engaged in initial training and in-service work, as well as teachers who are looking to innovate.

It is safe to say that no one can claim to be an expert in teaching. In this sense we are all in the process of learning and, as teachers become more practised in task setting, managing new techniques and encouraging active learning, it will all become easier and more familiar. This book is intended to ease the transition towards that stage.

Di Bentley and Mike Watts

ACKNOWLEDGEMENTS

We are most grateful to the contributors whose work provides the focus for the book. They represent but a small sample of the many talented and innovative science teachers in the country, who, with true professionalism, seek constantly to improve their own teaching and the learning experiences they provide for youngsters. We are particularly grateful to all our colleagues within the Secondary Science Curriculum Review for long and continuing discussions about teaching and learning in science which have helped in the shaping of our ideas.

1: LEARNING TO MAKE IT YOUR OWN

Mike Watts, Di Bentley and Joseph Hornsby

INTRODUCTION

We begin with a short story.

Gilbert Johns' Class

'Why do we have to learn about latent heat Sir?'

Harry's voice brought Gilbert Johns back to the reality of his fifth-year physics class. Given a choice, Gilbert would choose to teach 'A' level physics all the time because that is both his own specialism, and he enjoys his subject – especially at the level where the youngsters are beginning to get to grips with it. He looks around the room at the groups of children working, and knows that despite some difficulties now, the majority of the pupils in the room will get good results because he is careful to ensure that they all leave at the end of the lesson having 'got' the right idea about the topic. And, today, latent heat will be no exception. As he turns towards Harry's group to answer the question, he passes two other groups who are writing the method from their work card, a third drawing the diagram, and a fourth plotting the graph. He pauses to watch one group member reading the temperature of the water in the calorimeter and checks to make sure that Paramjit's labelling is correct, and that Susan is reading the thermometer accurately. He has a certain sense of satisfaction at the orderly way his pupils approach their experiments and feels that he is giving them – through an insistence on methodical working – an idea of the nature of science as a clear and orderly discipline. He sees his task as putting over the subject to children who have no understanding of science, getting across the concepts and facts about the way that the natural world works.

'Sir, why do we have to learn about latent heat, Sir?' Harry's voice breaks into his thoughts again.

'A good question', he thought, pausing to adjust the angle of the thermometer so Harry could see it better '. . . it deserves a good answer . . . because it will help you to understand why ice melts in your coke on a hot day? . . . no.'

'Because it's part of the natural world about us, Harry. Science helps us to understand these things and explain our world to ourselves. That's why it's on the syllabus. Don't worry, you'll understand it better by the time I've finished explaining the formula to you. You'll get it right in the exam.'

Gilbert turned to the board, and said 'OK now fifth years, when you've cleared away your apparatus, bring your books round the front so we can make sure everyone understands the right way to do the calculations.'

An entirely fictional cameo, though any resemblance to actual persons is certainly not accidental. All of us are familiar with features of chalk and talk and thermometers. Few of us are under any illusion that teaching, learning and schooling are long and sometimes complex processes. Although we present a fairly straightforward picture of Gilbert Johns' activities, and those of his students, we know that any classroom involves many kinds of complicated interrelationships. For instance, groups as well as individuals often have distinctive characteristics and no two groups of youngsters are ever the same. People have moods, and a group's mood can affect the way that pupils respond, how well the lesson goes and how well individuals within the group learn. The weather can produce its own moods and we surely all know occasions when a windy, leafy, autumn day can bring high spirits; rain and mist can dampen enthusiasm; the first snows of winter or a sudden thunderstorm can rapidly re-focus attention and concentration. But there are other, more predictable factors too. For example, the physical circumstances of the room, the availability of resources, the structure of the course, the approach adopted by the teacher, all contribute to the success of lessons. As our title suggests, in this book we are concerned with approaches to teaching and learning. Our experiences within the school system tell us that there are good teachers, good lessons and some excellent practice to be seen. Our intention in the book is to help make available and spread some of this good practice.

There is wide agreement about what schools should produce. An influential publication on teacher appraisal (Suffolk Education Authority, 1985), for instance, argues that there is broad agreement as to the effects that teachers should have:

pupils should be learning subject matter which consists of information, skills and attitudes; at the same time they should be helped to recognise their increasing competence, feel better about themselves as individuals, become better, more cooperative members of the community, develop more responsibility, increase problem-solving ability, prepare for the world of work and develop independence.

That is, schools should be producing effective, independent learners – young people who can organize their day-to-day affairs and manage to be informed and productive members of society.

While there may be some consensus about the product of education, there is much less consensus about the processes by which we arrive at it. The processes which schools use to 'help', 'develop' or 'prepare' young-sters are often diffuse and unclear. Although we have been at it a long time, the real nub of the problem is that educators know relatively little about the processes of learning. At first glance there is a temptation to think of learning as merely a combination of good motivation and recall. At the same time, we know it is not that simple. We know that motivation can be both engendered and lost, and that recall alone is seldom enough to ensure meaningful learning.

We also know that learning does not happen in a vacuum. On the whole, teachers spend much time and energy trying to get the right 'chemistry' for good learning: a pleasant learning environment, cooper-ation between teacher and taught, good management strategies, effective resources, and so on. No book can provide a complete recipe for success-ful learning or teaching. Our plan has been to draw together a wide range of classroom strategies through case studies by practising teachers; strategies which, in the past, the contributors have found have helped towards that chemistry.

PASSIVE AND ACTIVE LEARNING

We return in a moment to our cameo of Gilbert Johns' lesson, but begin first by making a distinction. Throughout this book we highlight the differences between 'passive' and 'active' learning. Our argument goes as follows:

1. Passive learning is the staple diet for many learners in numerous school classrooms.
2. Passive learning may suit some learners some of the time, but it is ineffective for many learners for much of the time.
3. Active learning means involving learners fully in their own learning, moving some of the responsibility for learning to the learner.

4. Encouraging active learning involves using different approaches to teaching.

We need to say, too, that we see two kinds of passive learning. You might not think, for example, that Gilbert Johns' group seems very passive. For instance, passive can – and too often does – mean rows of quiescent youngsters copying diagrams from a book or notes from the board. Gilbert Johns' group, on the other hand, seems to be actively engaged in practical work. But 'activity-based' learning and 'active' learning are different. Activity can be superficial and hide what is really quite passive learning. It is what is sometimes called 'busy-time' or, in the laboratory, 'recipe science' – classroom activity that has little basis in meaningful learning. What we mean is, that while there is practical work and experiments, it is not work that is initiated, designed or organized by the youngsters themselves. Their role is to follow step-by-step instructions, to copy the 'method' into their books, or fill in the blanks on a worksheet. In this sense they do not *own* the activities and often have little investment in either the process or the results. Towards the end of the lesson the learning is reinforced by making 'sure everyone understands' and has the right answers. In many respects we would also class this as passive learning. That is, what often passes for active learning is really passive learning in disguise. We leave our explanation of active learning for a little later and we return to it more fully in our final chapter.

In the writings about learning, passive learning is sometimes identified with a 'transmission' approach. This is a term borrowed from Douglas Barnes (1976) and Clive Carré (1981). The point both authors make is similar. For them, teaching can be loosely categorized into two general styles – the 'interpretative' and the 'transmission'. Carré identifies a teacher who follows a transmission mode as one who:

> believes knowledge to exist in the form of public disciplines which include content and criteria performance. This often means that they see themselves as 'authorities' in a subject;
> values the learner's performances insofar as they conform to the criteria of the discipline;
> sees the teacher's task to be the evaluation and correction of the learner's performance, according to criteria of which he/she is the guardian;
> sees the learner as an un-informed acolyte for whom access to knowledge will be difficult since he must qualify himself through tests of appropriate performance.

Maybe this is a caricature of the old-fashioned schoolmaster or mistress and is no longer a good picture of modern teaching. However, there is still some remnant of this style to be seen in the way teachers teach. Denis

Fox (1983), for example, casts a slightly different light and uses the term 'transfer theory' to refer to teachers within the transmission mode. He suggests that teachers who adopt the transfer theory see knowledge as a commodity which can be transferred, by the act of teaching, from one 'container' to another. Such people tend to express their view of teaching in terms of 'imparting knowledge', 'conveying information', 'giving the facts' or 'putting over ideas'. Of all the teaching methods, the lecture and the chalk-and-talk approach are the classical ways of seeing the transfer – or transmission – theory in action. In their study, Galton and Eggleston (1979) call science teachers who adopt this kind of approach 'informers'. Teaching like this means presenting an array of unequivocal facts in lessons, and directing pupils to sources for fact finding. In the lessons they observed this was mirrored by pupils referring constantly to teachers for the purpose of acquiring and confirming facts.

Teachers – and many pupils – who value a passive approach, then, have a particular view of knowledge and what it means for someone to acquire knowledge. Like other science teachers, our Gilbert Johns has views about the nature of knowledge in general, and teaching and learning in science in particular. The views that teachers have about knowledge and how it can be imparted and obtained, are obviously important for their teaching. In just the same way, science teachers' views of what science is, and how it works, are important for the development of young scientists. Both aspects together are extremely important for those concerned with changing the shape of the school science curriculum for the better. As Dick West (1986) says:

> Firstly in science education we really must take on board the whole question of what is knowledge and therefore what is science, and the linked question of ownership of knowledge. I don't feel we can plan and implement an effective science curriculum, or devise effective GCSE criteria, and validate and accredit courses until we have sorted this matter out.

We have associated our imaginary teacher with physics although Gilbert Johns is probably fairly typical of teachers of all 'brands' of science. It is our view that a transmissionist approach to teaching – and a passive mode of learning – predominates in science classrooms. But why shouldn't it?

WHY NOT ENCOURAGE PUPILS TO BE PASSIVE?

In his book, Clive Carré (1981) proposes five possible reasons why teachers use such an approach:

1. The most plausible argument is time, something we're very short of. If teacher talks and pupils listen, there are no 'red herrings' and the ground seems to be covered that much more quickly. [This has often been exacerbated by the demands of examinations which put considerable constraints on what happens in lessons.]
2. There's a strong influence from tertiary education, especially in science, for those teachers who have, in their own time, been recipients of bodies of knowledge handed over by lecturers for regurgitation at a later date: there's respectability in the transmission doctrine.
3. A clear message from many pupils indicates that they would prefer teachers to do all the work, and that chaos might develop where the tradition is broken by those teachers who may not wish to adopt a 'telling' strategy.
4. One way to develop confidence in using science language is through an apprenticeship system. The science teacher needs to give examples, often, of impersonal precise talk within the accepted framework of science expression. So there is a place for 'we set up', 'control all variables', 'it was allowed to', or 'there is a tendency to'. (All walks of life have their own phraseology – builders 'offer up' when fitting one piece of timber to another.)
5. A real concern may exist that the 'interpretative' stance may be dangerous. A transmission mode enables the teacher to have a tight hold of the reins and to know what's going on. Too much self-discovery, doing things on one's own initiative, could be breaking laboratory rules, and be a hazard to health.

Perhaps, too, as we mentioned earlier, the reason is a philosophical one and the answer lies in the deep-seated image that teachers have of knowledge and particularly of scientific knowledge. With very few exceptions, school science textbooks, and science teachers generally, project an image of science which has been described as empiricist-inductivist. Cawthron and Rowell (1978) and Arden Zylbersztajn (1983) are examples of writers who have stressed this point. Put very simply, empiricist-inductivism sees knowledge as there for the discovering and the taking. The problem it tackles is unearthing the right knowledge and putting it to use rather than constructing and interpreting it.

It is a way of thinking that has dominated science for many years and, as David Stenhouse (1984) points out, has had a strong influence on school science as it has developed. These may be very good reasons why it is such a dominant and widespread way of operating.

Carré's five reasons might explain why classroom activities are as they appear. However, there is a growing push for a shift in emphasis – a strong move for change.

SCIENCE EDUCATION POLICY

Over recent years, a considerable body of opinion has grown in favour of change. For example, the Department of Education and Science (1985) say:

> Although there are some impressive exceptions, too much of the time spent learning science by too many pupils consists of the accumulation of facts and principles which have little perceived, or indeed actual, relevance to their daily lives as young people or as adults.

Their evidence for that statement comes from the many HMI reports on schools and groups of schools. For instance, one recently published report says of the science classes seen that they suffered from:

> deficiencies in course planning . . . [and] . . . in the teaching styles and methodology. Even in schools where good practice was seen, there were examples of poor planning and enthusiasm for the subject. There was much over reliance on work cards and their use was not well planned . . . teachers talked for extended periods without recourse to class discussion. When discussion happened, it was often ill-disciplined and aimless. . . . Pupils were rarely given the chance to plan their own investigations or even asked to suggest experiments. (HMI, 1987)

That is, in the view of school inspectors and state policy makers, there is too much passive learning, too much transmission teaching. Their opinions are based upon many visits to schools and a view of the way science education should be progressing. However much there is a move toward greater centralization of the curriculum, it is often easy to dismiss the words of civil servants and politicians as being unrelated to everyday classroom needs. But in this case they are not alone in their views of science education. For many years there has been widespread pressure for change in school science, coming not least of all from the Association for Science Education (ASE, 1979, 1981).

The message relates to the need for a broad, balanced science curriculum for all pupils in the 11–16 age range. Curriculum guidelines show an emphasis away from traditional teaching and towards problem-solving approaches, increasing technological awareness and competence, and linking science education with social, cultural and personal issues. The Department of Education and Science, for example, believe that adopting different teaching approaches to those currently used would be 'conducive to the success of the national policy objectives' defined in their policy statement (1985). Changes are necessary, it is argued, in order to encompass the ideals of broad and balanced science for all. They go on to say, in a somewhat long – but telling – quote, that they believe that:

> science teachers have much to gain from considering the implications for science of the variety of methods defined in paragraph 243 of the Cockcroft Report as desirable means for the more effective teaching of mathematics:
>
> > problem-solving;
> > investigation;
> > practical work;
> > exposition;
> > discussion; and
> > consolidation and practice.
>
> Such a range of teaching approaches seems to the Secretaries of State to be essential for the teaching of science as well as of mathematics; the key to success lies in flexibility and variety within such a repertoire. In particular, opportunities for pupils to contribute their own ideas to discussion are important, with the object of establishing that in science, recourse to experiment and experimental data is the principal means of testing whether a hypothesis is supported by evidence, and if so, how far its implications extend. The opportunities for pupils to engage in experimental work in which a variety of practical and investigative skills are developed under the supervision of the teacher is also of crucial importance. The balance in practical work should be more towards solving problems and less towards illustrating previously taught theory.

The reference to the Cockcroft Report in the early part of the quote is somewhat misleading. Cockcroft (DES, 1982) does indeed exhort mathematics teachers to consider a wide range of teaching approaches. The report, however, concedes in the very next paragraph that few mathematics teachers actually make use of such methods, nor is there very much research in mathematics education to support such moves. Other educationalists represent similar views and these have been documented elsewhere (see e.g. Ingle and Jennings, 1981).

Two important influences worth noting separately come from the National Curriculum, and what has been called the 'new education' (Ranson *et al.*, 1986). The latter relates to Technical and Vocational Initiatives (TVEI) and the extension of this (TVEE).

Many in education fear that the National Curriculum will deaden innovation and curtail creative development. Only time will tell. What practice is like in schools can only be evaluated after a number of years of implementation. The rhetoric would not lead one immediately to fear the worst. For example, the Science Working Group (DES, 1987) say of the science programme, that:

> it is assumed that schools and teachers will interpret the programme so as to provide learning experiences which develop pupils' science understanding, skills and attitudes in ways which stimulate interest and help pupils see the relevance of their studies for their everyday lives. . . . The group has attempted to keep the areas of study to a minimum, in order for pupils to be taught in ways which promote their confidence in learning and encourage enjoyment and interest in science.

In such directions lie some hope of innovative approaches to teaching. While the 'new learning' is not exclusively in science and technology, these two aspects do form the power-house that drive it. Its major features have been listed by Hannon (1986) as:

> reform of the curriculum (modular organisation, shifts towards integration rather than subject divisions);
>
> experiential learning, especially through off-site activities and practical experience in the community;
>
> pedagogic shifts towards participative group work, away from didactic methods;
>
> reform of assessment: credit accumulation, profiling etc;
>
> vocationalism – either in the narrow sense of preparation for work, or in the broader sense of taking the world of work as an important site of study and experience. Work experience is a key component here;
>
> parental – and pupil – involvement: stress on negotiation and choice.

We would also suggest a strong emphasis upon problem solving, however that term is to be defined. As Ranson *et al.* (1986) say:

> The new education pursues a policy of whole curriculum planning for whole people and seeks to weaken the boundaries between areas of experience so as to integrate the curriculum. A broad curriculum is advanced which embraces cognitive skills, of knowing but also problem solving; experience in the community (celebrating, perhaps, public service) and in work; and giving greater emphasis to creativity in the arts but also social relationships. . . . The new education places considerable emphasis upon the transforming of teaching methods if achievement is effectively to be extended for all young people. Formal didactic exposition of factual knowledge is not calculated to kindle interest.

The important point about all this is that there is a strong mood for change among those who advise, inspect, make policy, support and use the products of school science.

CHANGES FROM WITHIN SCIENCE EDUCATION

'If I knew where jazz was going, I'd be there already', Humphrey Lyttleton is rumoured to have said. Many people relish being at the forefront of change. Science teachers have, in the past, been embroiled in a series of curriculum development schemes, developed by science educators to try to meet the needs of science education. There was Nuffield in the 1960s and 1970s, along with LAMP, SCISP (Schools Council Integrated Science Project) and Science at Work. In the 1970s and 1980s there has been Nuffield 13–16, SISCON (Science in a Social Context), SATIS (Science and Technology in Society) and a wide number of schemes that have grown through the auspices of the Secondary Science Curriculum Review (SSCR). Established in 1981, the overall aim of the SSCR has been to stimulate the development of science studies for all young people whatever their career intentions. The Review has been very successful in enabling science teachers to review and develop their own school science curriculum, and also its content, purposes, resources and teaching methods. As they say of their own work (SSCR, 1986):

> During its initial five-year period it roused such wide interest that the resources developed far exceeded expectation and some eighty local education authorities and 270 working groups of teachers, advisers, lecturers and representatives of industry and commerce were involved in some aspect of its development work.

The SSCR is a good example of teachers taking the time and opportunity to reflect upon the daily tasks of teaching and learning in classrooms. Positive change is usually the creative outcome of teachers sorting out some of their persistent problems.

EVIDENCE FROM RESEARCH IN SCIENCE EDUCATION

Another telling argument against passive learning and transmission teaching comes from studies of classroom learning and performance in science. Put simply, it may work for some youngsters for some of the time, but it is not very effective for many youngsters for much of the time: some people find the transmission mode of teaching to be somewhat less than successful. Galton and Eggleston (1979) make the point that, compared to both a more interactive mode and a 'problem-solving' approach to teaching science, the 'informer' method was the most commonly used and yet was reported to be the least effective. Similarly, findings from the Assessment of Performance Unit (APU) have given the community of science educators good reasons to doubt the effectiveness of traditional teaching methods and routine (passive) practical work. The APU (1986), for instance, ask the question:

> Could it be argued that many routine school science experiments appear to make too few procedural demands of many pupils, and that much more could reasonably be expected of them in terms of experimental design and the collection and interpretation of data? With a few provisos regarding poor performance on a few aspects, could it be that children's ability at this level is frequently underestimated, and that they could be much more actively involved in the design and carrying out of investigations, in the making and interpretation of observations etc. than is perhaps the case at present?

By way of answer the authors consider the performance of 15-year-olds both in practical work and in the application of basic scientific concepts, such as force and gravity:

> Certainly, each concept area so far described in any detail shows that while 15 year olds attempt to use commonsense and 'everyday' science knowledge to answer assessment items, a very low proportion of pupils demonstrate the kind of conceptual understanding towards which much routine school work is aimed.

We take from this that routine school work (recipe-science) is not particularly effective in promoting conceptual understanding – except for a very small proportion of pupils. The reference here to commonsense or 'everyday' science is an acknowledgement of a considerable body of research that has been conducted into children's learning in science.

There is now a growing number of publications which review this work: Driver and Easley (1978), Gilbert and Watts (1983), Driver and Erickson (1983), Bell *et al.* (1985), Driver *et al.* (1985) and Osborne and Freyberg (1985) are all useful examples. This research shows that youngsters bring to science lessons views of the world and meanings for words which have a significant impact on their learning. Children's learning in science is sharply influenced by what they already know. They often consider their own views to be more sensible and useful than those presented by teachers. As a result there are commonly major differences between teachers' intentions and expectations and the learning that actually takes place. In their study, Osborne and Freyberg (1985) summarize these differences as follows:

there was a disparity between the ideas children brought to the
    lesson and the ideas the teacher assumed that they would bring;
there was a disparity between the scientific problem the teacher
    would have liked the children to investigate and what they took to
    be the problem;
there was a disparity between the activity proposed by the teacher
    and the activity undertaken by the children, despite considerable
    teacher intervention;
there was a disparity between the children's conclusions and the
    conclusions proposed by the teacher.

That is, it has to be recognized and accepted that 'telling' by itself does not
always (can seldom?) produce the desired results. Teachers' understandings and conclusions cannot be transferred directly into pupils' understanding and conclusions. It is a recognition that young people, as do all
people, construct their own meanings from what others say and do.
These personal constructions embody their *own* intentions, expectations
and conclusions. Science lessons are frequently seen as isolated events
even though science teachers see them as a progression of linked topics
and ideas. To combat some of these problems, the science teacher's job
becomes that of:

encouraging the pupil to share and eventually own the purpose for
    the lesson or activity;
developing learning experiences that allow pupils to take responsi-
    bility for the design, process and outcomes of the investigation;
valuing pupils' hypotheses and conclusions and generating dis-
    cussion of the scientific description of what has been taking place
    in the activity.

This is a difficult task. As Osborne and Freyberg say:

Effective implementation of (these) ideas . . . requires active
teaching by a teacher who clearly appreciates children's ideas, the
scientific view to be encouraged, the types of activities which might
achieve conceptual change and the associated interactive teaching
sequences which need to be adopted.

That is, we need to provide more opportunities for pupils to talk about
what they are doing, to become aware of their own ideas and those
of their peers, and to modify their own ideas where necessary. It is not a
task or job description that is familiar to many science teachers, and it is
certainly not easy:

How can teachers carry out what they believe to be their responsibilities when these include control of pupils' learning and encouraging pupils actively to formulate knowledge? In one direction lies

control so strong that school knowledge remains alien to the learner (whether [s]he rejects it or goes along with it); in the other direction lies a withdrawal of guidance, so that learners never need to grapple with alternative ways of thinking. The teacher has to find [her] his way between the two. (Barnes, 1976)

THE EFFECT OF EXAMINATION CHANGES

It would be improper to consider all the other changes and moves of opinion in science education, and education generally, without mentioning the changing pattern of examinations and assessment. Parallel to the growth of curriculum pressure – and possibly as a direct result of it – new and more appropriate forms of assessment and evaluation have been and are being developed. These are exemplified in moves towards attainment targets for the National Curriculum [as reported by the Task Group on Assessment and Testing (DES, 1988)], in GCSE, in graded testing, pupil profiles and records of achievement, all of which in turn carry over their own demands. In many ways, GCSE *should* mean that we have the opportunity to explore new and different ways of assessing and examining science courses.

The advent of course-work, for example, means that the way is open to credit youngsters' work over a range of activities, rather than just through written examinations. Course-work assessment is a central feature of GCSE. Under the old system, course-work was sometimes a feature of both CSE and 'O' level: much good and valuable work was achieved, for example, in 'O' level practical work and CSE mode 3 projects. However, some element of course-work is now compulsory for all pupils. Moreover, it is a guiding principle of the examination that all candidates must be presented with tasks which they find manageable, satisfying and through which they can display positive achievement. That is, classroom tasks must either be closely matched to learners' abilities and competencies, or general tasks set which are then differentiated by outcome – by what pupils actually do or how they perform. In practice, this combination of requirements can make course-work highly interesting and rewarding for the learner. It can also present difficulties of organization and management for the teacher. For a useful critique of course work in science, see Howlett (1987), Bentley (1989) and Billing (1988).

We do not tackle either course-work or other aspects of assessment and examining directly in this book. Our emphasis is upon collecting together different approaches to teaching and learning – there are a growing number of books dedicated to discussing the implications of the examination system (we list some of these in the bibliography). Some of the contributors to the chapters that follow do mention aspects of examinations, and we leave it to them to discuss such issues as they arise in the context of classroom practice.

SO WHAT IS ACTIVE LEARNING?

Having dealt at length with passive learning, we need to attempt a description of active learning. We set out our points under two headings – 'Active learners' and 'Active learning needs'. The former lists some of the attributes of active learners, whereas the latter lists some of the requirements needed for active learning to take place. At the end of each chapter that follows we return to some part of these lists to explore items we think have been exemplified in the case studies.

Active learners

1. *Initiate their own activities and take responsibility for their own learning.* By this we mean that the task in hand, or the scheme of work, is something *they* want to do. It often comes from within them, from their need to know or to find a solution. It may be that the impetus or suggestion for it comes from the teacher or from outside the classroom. Nevertheless, they want to shape it themselves so that it becomes their task and they become accountable for its outcomes. This way they feel in control and fully involved in their own learning.

2. *Make decisions and solve problems.* We believe that active learners recognize the demands of particular tasks, take responsible decisions and seek ways to solve problems within them. They judge the task for what it is worth, and tackle it appropriately even when it derives from outside, from a scheme of work, from the teacher or some other source. Making decisions is important – only when learners make decisions towards the solution of a problem do they begin to own the problem for themselves.

3. *Transfer skills and learning from one context to other different contexts.* Active learners have feelings of ownership over information, data, interpretations and understandings, which means that they can judge the worth of facts and opinions. For us, ownership is important because it indicates that the learning taking place is (as Strike and Posner, 1985, suggest) intelligible, credible, fruitful and relevant to the youngsters concerned.

4. *Organize themselves and organize others.* For us, active learning means being able to work independently *and* work within a group. It is an important attribute of active learners that they know when a task is one to

be tackled alone or one that requires collaboration with others. Working individually, or working closely with others in a small group, does involve particular skills and abilities. Such skills enable them to become cooperative members of the community. They are also aware of the time requirements of different tasks and are capable of pacing themselves to meet deadlines. They use a range of study skills, and select the most appropriate resources and information and the means of gaining access to them.

5. *Display their understanding and competence in a number of different ways.* This means that youngsters select the most appropriate means of reporting their progress, what they know and understand. In discussions they are able to communicate and explain their ideas and understandings so that others can appreciate them. In doing so they tailor their report to match their audience showing an awareness of what style is appropriate.

6. *Engage in self- and peer-evaluation.* Active learners are effective learners. That is, they are confident enough to develop their own criteria, evaluate their own progress regularly and recognize their own competence and weaknesses. They are prepared to share these criteria and evaluations of their progress with their peers and teachers. They are willing to discuss them, defend them and where necessary review them in the light of others' opinions. They are also prepared to assist in the evaluation of others' progress and share those evaluations with the individuals concerned in a non-threatening and supportive way.

7. *Feel good about themselves as learners.* Active learners believe in themselves and grow in enthusiasm for what they are doing. They understand that learning is an emotional business, involving excitement, disappointment, sudden 'eureka' moments and periods of perseverance. Because they are engaged in establishing their own directions and progress, they are more likely to divine success in what they do. Success breeds confidence and, in turn, confidence breeds positive feelings and motivation.

Active learning needs

1. *A non-threatening learning environment.* Active learners are involved in speculation, experimentation and reformulation of their ideas and existing concepts. Such activities require emotional investments. The process of sharing and reforming ideas – admitting that one has not understood something thoroughly, for example – is often quite difficult. It can sometimes be accompanied by ridicule from either peers and/or the teacher. Youngsters may then be reluctant to engage with the whole

process. This means that the environment in which youngsters propose and test out ideas needs to be supportive, while still giving an honest evaluation of their efforts. They need to grow and change through this evaluation, yet also need to be sure that such evaluation will concentrate upon their 'professional' contribution, not demean them as individuals.

2. *Pupil involvement in the organization of the learning process.* Early involvement in the way the lesson, scheme or tasks are to be organized maximizes the possibilities of active learning. It allows learners to begin to stamp their own direction on what is taking place, orientate the activities to their own needs and influence events so that they feel it has some purpose for them. In this, it allows teachers and pupils to establish some common goals about how learning is to take place.

3. *Opportunities for learners to take decisions about the content of their own learning.* Growth towards independent autonomous learning means giving pupils opportunities to choose what they learn and how they display their learning to others. They need also to be able to evaluate their own learning, and decide about further directions in which they wish to move.

4. *Direct skill teaching.* Many of the attributes of active learning need to be taught directly and tutored. For example, while it is important to provide opportunities for participative group work, some of the skills of cooperation and negotiation involved in working in a group need to be taught directly. This is also the case for some study skills, as well as particular scientific skills.

5. *Continuous assessment and evaluation.* For learners to develop a realistic sense of their own worth, and the value of their ideas, they must be involved in the evaluation of those ideas as they progress. Such an involvement enables them to diagnose their strengths and weaknesses, and take their own steps to build on or rectify them.

6. *Relevance and vocationalism.* In our view, active learning needs to be relevant learning. It may well be possible to use alternative approaches in the classroom; for example, to have youngsters memorize an obscure set of nonsense syllables in an 'active' way, play 'Trivial Pursuit' in order to accumulate facts on 'science and nature', or enact charades of famous personalities in chemistry. However, alternative methods alone are not sufficient to engender active learning. Such methods only really become active when they have a purpose, i.e. when they achieve relevance by being set within the context of everyday life and the world of work and leisure. Our intention is not to endorse narrow vocationalism, an ethos of producing 'cannon fodder for industry', but to see learning in the broader sense of exploring the ways of the world. And within that, to take the world of work as an important site of study and experience.

THE CHAPTERS OF THIS BOOK

This chapter, then, has dealt with some of the problems concerned with science in education and with teaching and learning in science – the solution of which requires some change on behalf of the teacher. We hope the discussion so far has suggested some answers to the question 'Why should I change the way I teach?'

This book is about change: as we have discussed there are many new ideas being developed both inside and outside schools. In our view educational change can be seen in two ways. First, it can be thought of as somewhat random, a series of policies and practices that, at best, deserve the label 'trial and error'. Where change does bring about improvement, it is often the result of an accumulation of many modifications along the way – some for the worse, though, on average, more for the better. Secondly, change can be seen as the product of rational goal-directed activity, as the best choice among a set of feasible innovations designed to solve a series of problems. The cynics among us will shy away from change, fearing too much of the first; idealists will welcome change for the chance of taking part in the second.

In this book we tread a wary path between the two. We invite teachers to adopt differing approaches to teaching and learning, and to do so by choosing and planning for them carefully. The main part of the book consists of contributions which cover a range of styles and techniques in an attempt to offer one set of feasible and workable innovations designed to tackle particular problems. They are not, however, fail-proof recipes. Trying something new means there will inevitably be some element of trial and error. Mistakes will be made and, of course, much learning will take place through experience. In spite of this, we believe that overall the changes will be for the better. There is increasing recognition that *any* change will need a quite radical review of the methods of classroom teaching that are required. Many teachers are unaware of different styles of teaching, their potentials and benefits, or have yet to engineer the opportunity to try new ways. Some will therefore need an introduction to the variety of methods available, their point and purpose, and some rationale for their use. Others will be aware of methods used by colleagues elsewhere in the educational system, yet may have lacked the confidence to try them out for themselves.

The contributions in the book vary enormously. We have clustered them together in chapters according to the primary focus of what is described, be it educational drama and role play or problem-solving. There is no reason, however, why youngsters cannot solve some problems through role play, and so the strategies can sometimes be seen to merge into each other. Despite the variation, the strategies do have a number of common threads. We have also tried hard to 'balance' the contributions in a number of ways.

First, they are all embedded in science education, though not all the writers are confirmed scientists – some come from other parts of the educational system. We have tried to develop a wide agenda, to embrace a broad view of science that includes, for example, aspects of health education, the social implications of science, science and technology, as well as more traditional sciences.

Secondly, all of the contributions develop ideas that are a little out of the ordinary. This gives rise to one of our meanings for the term 'alternative' in our title. It may be the subject matter that is unusual, as in teaching about biotechnology or AIDS. It may be the classroom activities that are different, as in playing card games about electric motors, or role playing parts of Galileo's life. We have tried to balance the more extraordinary contributions with some examples of good ideas within more ordinary settings.

Thirdly, and perhaps most importantly, each strategy – in its own way – hands over some part of the responsibility for learning to the learner. They all, in large part, encourage pupil participation and promote active learning and, in some small way, attempt to engender independence and autonomy. Again, we have tried to balance examples where the youngsters are invited to take a large measure of responsibility for their own learning with others where teacher guidance is stronger.

Finally, they are rooted in actual classroom practice and so deserve the title 'practical' alternatives. We have asked each contributor to write to a particular brief. They have had the difficult task of portraying their example of good practice so that their teaching situation could be recognizable, of including as much detail as possible of pupil and teacher organization, and of noting the success and failures of various activities.

On the whole, we address the book to individual teachers to help in the planning and development of their day-to-day work. Clearly, though, no teacher works in isolation and the development of new approaches may well depend upon the organization of courses and, therefore, be contingent upon group decisions or departmental organization. Hopefully, there is something of value here for groups and department teams, as well as the lone class teacher. There is certainly something to be said for all specialists being aware of the work being carried out in other subject areas. As the boundaries between subjects become increasingly diffuse, it behoves all of us to be informed of the developments in neighbouring subjects, and of their approaches to learning and teaching.

BIBLIOGRAPHY

Assessment of Performance Unit (1986). *Science in Schools; Age 15*. Report No. 4, Department of Education and Science. London: HMSO.

Association for Science Education (1979). *Alternatives for Science Education: Consultative Document*. Hatfield: ASE.

Association for Science Education (1981). *Education Through Science*. Hatfield: ASE.

Barnes, D. (1976). *From Communication to Curriculum*. Harmondsworth: Penguin.

Bell, B., Watts, D. M. and Ellington, K. 1985. *Learning, Doing and Understanding: The Proceedings of an SSCR Conference*. London: SSCR.

Bentley, D. (1989). *GCSE Coursework: Science. A teacher's guide to organisation and assessment*. GCSE Coursework Series (Eds J. Nixon and M. Watts). London: Macmillan.

Billing, P. (1988). *GCSE Coursework: Biology. A teachers' guide to organisation and assessment*. GCSE Coursework Series (Eds J. Nixon and M. Watts). London: Macmillan.

Carré, C. (1981). *Language Teaching and Learning: Science*. London: Ward Lock.

Cawthorn, E. R. and Rowell, J. A. (1978). Epistemology and science education. *Studies in Science Education* **5**, 31–59.

Department of Education and Science (1982). *Mathematics Counts. The Report of the Committee of Inquiry into the Teaching of Mathematics in Schools*. London: HMSO.

Department of Education and Science (1985). *Science Education 5–16: A Statement of Policy*. London: HMSO.

Department of Education and Science (1987). *National Curriculum Science Working Group*, Interim Report. London: DES.

Department of Education and Science (1988). *National Curriculum Task Group for Assessment and Testing. Interim Report*. London: HMSO.

Driver, R., Guesne, E. and Tiberghien, A. (1985). *Children's Ideas in Science*. Open University Press: Milton Keynes.

Driver, R. and Easley, J. (1978). Pupils and paradigms: A review of literature related to concept development in adolescent science students. *Studies in Science Education* **5**, 61–84.

Driver, R. and Erickson, G. L. (1983). Theories-in-action: Some theoretical and empirical issues in the study of students' conceptual frameworks in science. *Studies in Science Education* **10**, 37–60.

Fox, D. (1983). Personal theories of teaching. *Studies in Higher Education* **8** (2), 151–63.

Galton, M. and Eggleston, J. (1979). Some characteristics of effective science teaching. *European Journal of Science Education* **1** (1), 75–85.

Gilbert, J. K. and Watts, D. M. (1983). Concepts, misconceptions and alternative conceptions: changing perspectives in science education. *Studies in Science Education* **10**, 61–91.

Hannon, V. (1986). The 'new education': what's in it for girls? In *The Revolution in Education and Training* (Eds S. Ranson, B. Taylor and T. Brighouse). Harlow: Longman.

Her Majesty's Inspectorate (1987). *Report by HM Inspectors on Survey of Science in Years 1–3 of some Secondary Schools in Greenwich*. London: DES.

Howlett, A. (1987). *GCSE Coursework: Physics. A teachers' guide to organisation and assessment*. GCSE Coursework Series (Eds J. Nixon and M. Watts). London: Macmillan.

Ingle, R. and Jennings, A. (1981). *Science in Schools: Which Way Now?* London: NFER-Nelson.

Osborne, R. and Freyberg, P. (1985). *Learning in Science*. London: Heinemann.

Ranson, S., Taylor, B. and Brighouse, T. (Eds) (1986). *The Revolution in Education and Training*. Harlow: Longman.

Schools Council Integrated Science Project (1974). London: Schools Council and Longmans.

Science at Work Series (1980). Taylor J. (director). London: Addison Wesley.

Secondary Science Curriculum Review (1986). *Better Science: A Directory of Resources*. London: Heineman and ASE.

Stenhouse, D. (1984). *Active Philosophy in Education and Science*. London: George Allen and Unwin.

Strike, K. A. and Posner, G. J. (1985). A conceptual change view of learning and understanding. In *Cognitive Structure and Conceptual Change* (Eds L. H. T. West and A. L. Pines). London: Academic Press.

Suffolk Education Authority (1985). *Those Having Torches . . . Teachers Appraisal: A study*. London: HMSO.

West, R. W. (1986). Coping with the challenge of change. Presidential address to the Association of Science Education, London Region, October.

Zylbersztajn, A. (1983). A conceptual framework for science education: investigating curricular materials and classroom interactions in secondary school physics. Unpublished Ph.D. thesis, University of Surrey, Guildford.

2: PRACTICALS AND PROJECTS

INTRODUCTION

The influence of the Nuffield curriculum development project in the 1960s and 1970s has meant that practical investigations are a feature of most science classrooms today. Certainly, lower-school science classes involve a high percentage of practical investigation. Beatty and Woolnough (1982) surveyed 11–13 science and found that youngsters spend between 40 and 80% of their time doing practical work. Such findings do, however, prompt the following questions: What *is* practical work in science, what are its aims and purposes and what do youngsters actually do? Are they gainfully employed? Or is it simply another set of 'busy time' activities? For the teachers in the Beatty and Woolnough survey, practical work had many aims, the five most important of which were to:

encourage accurate observation and description;
arouse and maintain interest;
promote a logical reasoning method of thought;
make phenomena more real through experience;
be able to comprehend and carry out instructions.

What did youngsters actually do to achieve these aims? The teachers described four types of activity: 'standard exercises' (emphasizing particular procedures and developing skills in using them), 'teacher-directed discovery experiments', 'demonstrations' and 'project work'. Of these, the first two were the most common and the last the least common.

Summarizing this research, practical activities in lower-school science appear to have two major purposes: to develop practical skills and particular attitudes. In our opening chapter, we referred to a fictional lesson in which youngsters were involved in practical work and yet, in our opinion, were not engaged in active learning. That is, the tasks were not their own, and did not necessarily enhance their understanding of science. They were essentially 'teachers' tasks'. This kind of situation is not entirely fictional. For instance, Her Majesty's Inspectorate have reported that:

> Practical work is undertaken but the work gives few opportunities for pupils to design investigations or interpret observations. The attention given to teaching practical skills is spasmodic and insufficient. Pupils start their courses with high expectations; they enjoy working in specialist rooms and being involved in practical work. . . . However, the pupils seen were often more bewildered than enlightened by what they did. They were generally not challenged sufficiently by the work. (HMI, 1987)

Clearly, HMI saw little by way of active learning. If this seems over-critical, we note that none of the teachers in the Beatty and Woolnough study highlighted features we would consider to be part of active learning. Their description of youngsters' normal activities in class was 'teacher-directed discovery experiments'.

WHAT MIGHT PRACTICAL WORK IN SCHOOL SCIENCE BE?

There is some agreement. In *Science Education 5–16*, the DES (1985) is quite clear about the purpose of practical activities in science:

> The Secretaries of State attach great importance to encouraging effective learning through extensive practical experience. In the first three secondary years the experience of scientific activity is still new and exciting to many pupils. Skilled teaching, well matched to the individual abilities of pupils can capitalise upon that sense of excitement and produce continuing enthusiasm for science.

The purpose of practical experiences in the lower secondary school, then, is to encourage enthusiasm and excitement. This too is one of the most important aims for teachers. The DES, however, does criticize present practice as fulfilling the function of 'illustrating previously taught theory' rather than encouraging skills. From our point of view, we thoroughly endorse their later comment, one we have already quoted, that:

> opportunities for pupils to contribute their own ideas to discussion are important, with the object of establishing that in science, recourse to experiment and experimental data is the principal means of testing whether a hypothesis is supported by evidence, and if so how far its implications extend. Pupils should be given the opportunity to engage in experimental work in which a variety of practical and investigative skills are developed.

Practical work, then, is being seen as a way to encourage skill development and enable youngsters to test out their ideas about science. In this sense GCSE course-work assessments can have a profound effect on practical work in both the lower- and upper-school. Observation, hypothesis testing and following instructions will be highlighted in years 4 and 5 as the tendency is at present in lower-school science. Hopefully, too, more active learning features (such as designing experiments) will also become commonplace.

Our four case studies illustrate some of these points. The first two, by John Heaney and Mary Doherty, explore practical work through projects. Secondary school teachers use projects very infrequently: middle-school teachers make plentiful use of them. Perhaps this indicates the influence of primary schools, where project work is a highly developed way of working. As both Heaney and Doherty point out, projects allow pupils to shape the direction of their learning for themselves, be in greater control of the time they want to spend on different aspects and follow interesting ideas and avenues as they arise. Mary Doherty also suggests that projects allow youngsters to develop confidence in their ability to use their science skills effectively. She discusses how projects might be used as a means of providing assessment materials for GCSE course-work. We regard both cases as displaying a high degree of active learning. Philip Naylor's study continues the theme. As he says:

> I have become increasingly committed to practical work as a means of pupils developing skills, but more importantly it is a purposeful way of learning about the world.

He describes how particular aspects of a fourth-year science syllabus are made more meaningful when youngsters locate the work in their own experience and take responsibility for identifying and developing different aspects of a whole problem.

In the last example of practical work Di Bentley describes helping lower-school pupils develop skills through scientific processes. She focuses on the use of real-life situations and invites pupils to explore different scientific solutions to the problems faced by the people in these circumstances.

CASE STUDY 1: PROJECT WORK

John Heaney

By project work I mean any piece of work done by a student or group of students in which they exercise a considerable degree of autonomy and responsibility. It is a form of active learning. It can be an investigation or a 'design and make' exercise. The theme or context is agreed between teacher and students. Each student is expected to produce some tangible end-product, a report or maybe an artefact, and where it has been a group venture, it should be clear what each individual has contributed. A simple review of a topic culled from the literature is not usually very fruitful because of the temptation to copy or paraphrase without any really creative involvement by the student. However, reading in association with a practical investigation, together with pertinent reference and acknowledgement, is certainly to be encouraged. Sometimes, especially with younger children, a whole class can contribute to a project, but in this paper I am mainly concerned with secondary students aged 14–16 working individually or in small groups.

The aim of project work is to develop scientific skills and processes and to encourage desirable attitudes. The processes might include retrieving and processing information, designing and carrying out experiments, drawing conclusions, problem solving, formulating and testing hypotheses, etc. Good project work should also engender qualities which are perhaps even more valuable, e.g. personal responsibility, self-criticism, planning and organization, patience and perseverance, cooperation with others. Ultimately, the teacher's role should be to counsel and support rather than to direct. This may not be very realistic in the early stages, except perhaps for very able students, but the aim should be for all students eventually to become independent, choosing their own topic, formulating an approach, selecting and carrying out background reading and research, executing and interpreting experimental work and writing reports. Very few will achieve this without much practice and plenty of support. The teacher's job, despite its low profile, requires a great deal of hard work as well as sensitivity and forbearance.

Successful project work demands both structure and flexibility and these two requirements are entirely compatible with each other. Experienced students can structure their own projects but novices will need to have a structure provided, and probably a lot of direction too. How can this be provided within a flexible, student-centred approach? How can responsibility for planning, decision making and detailed organization gradually be transferred from teacher to student? How can differentiation be achieved?

The approach described below has worked with mixed-ability groups of up to 24 students. The course was one of the CSE science options and occurred at several places in the option pattern. I think it could also function well as part of a core science, blocked time-tabled structure within which a team of teachers could pool their expertise and share the workload. Each student did at least two 'special studies', which together counted a maximum of 25% toward their CSE (Mode 3) mark. The final project counted a further 30%. They have enjoyed it and many of them achieved good results. It should be possible to negotiate a similar procedure for allocating course-work marks in GCSE courses. The teachers' workload was heavy but manageable. Laboratory technicians were kept busy too.

Before they began 'real' projects the students tackled a series of 'special studies'. The basic textual material was a set of 'Introduction and Assignment Sheets'. We started with six but gradually increased our collection to reflect the interests and expertise of the various teachers in the team. Here is one early example which, like the others, was illustrated by a collage of photographs and diagrams. It was based on *Nuffield Secondary Science*, Field of Study 8.3. It contained about 6 weeks' work but some students did not do all the experiments in this time. None of the introduction and assignment sheets occupied more than two A4 sides.

Introduction and Assignment Sheet 6

THE WEATHER

We live at the bottom of an ocean of air. Strange and exciting things happen in this ocean – rain and wind, heat and cold – weather. In this unit you can study what causes the weather and perhaps even learn to forecast. But don't expect always to get it right; even the experts can't manage that, although they don't really do too badly. Our atmosphere is so complex, so restless and changeable, that we must expect it to surprise us every now and then!

These are some of the questions that you might like to explore.

What are clouds made of?
Why don't clouds fall down?
What causes rain?
What is wind?
Why does the wind change direction and strength?
What causes thunder and lightning?

EXPERIMENTS

0601 Does air have mass?
0602 The composition of the atmosphere
0603 Moisture in the atmosphere
0604 Clouds
0605 How does heat make the air move?
0606 Warming up sand, soil and water
0607 Land and sea breezes
0608 Moving air masses and fronts
0609 Weather forecasting
0610 Weather satellites

The sheets served two purposes. First, they enabled students to get an idea of what each option entailed and so to make a reasonably well informed choice. A complete set was displayed on the laboratory wall. Selection was also helped by students having previously watched and talked to each other about their work, because it was quite normal for as many as five different options to be running in the same class at any one time. This made life hectic for the teacher but did relieve the pressure on scarce resources. Secondly, the sheets provided outline schedules of work. Associated with each one was a set of cards for the technician which listed the apparatus and materials required for each experiment. The numbers (0601, etc.) are accession numbers for this card index. Special apparatus was kept boxed and labelled with the number.

The students negotiated their choice with the teacher. Usually they got their first choice, but if an option was very popular some were persuaded to postpone it to a later date. Teachers also guided students toward particular options which seemed appropriate to their needs – but the choice was ultimately the student's (and had to be seen to be so).

At the end of each lesson, students were responsible for ordering the experiment they needed next time. This was done on a single sheet of paper for each class, with students' names listed and a column for each period. Everybody simply filled in an experiment accession number under the appropriate date. They soon learned how much trouble it caused both themselves and others if they wrote in the wrong number or forgot to fill it in altogether. Absentees just got next time what they had requested for the lesson they missed. Some students asked for the same experiment on successive occasions or, perhaps, usually on the advice of their teacher, changed the order of the experiments, or left one or more out. Both teacher and technician used the list to prepare materials. It also provided the teacher at a glance with a quick record of a student's progress.

Preparation by the teacher for each lesson was individualized and this is where the flexibility arose. Standard worksheets – sometimes more than one alternative – were available for each experiment and kept in a file. If one was appropriate the student's name was written at the top ready for handing out at the beginning of the lesson. Alternatively, the student was referred to a book or some other resource. Sometimes the teacher decided to write a worksheet especially for an individual. It may have been highly structured or very open-ended, lengthy or very brief (perhaps a single sentence), directive, discursive or questioning – in short, exactly what the teacher felt the student needed at that stage. The experiments were reference points but did not limit the type of work done or the nature of the communication between teacher and student. The latter included background reading, leading questions and indeed a whole range of stimuli. It was often appropriate to address the communication to a group of students but each individual got a copy. (The photocopier was kept busy!) Any new resource material was stored in the file for future reference and possible use.

This procedure enables progression. It requires great self-discipline by teachers especially if several of them are using the same file. A formative assessment of each student has to be made, formally or informally, before each lesson, so the teacher really keeps in touch. The actual workload, once the teacher is at home with the procedure, is no greater than traditional marking and far more interesting. Each student becomes engaged in a disciplined dialogue, partly spoken and partly written, with his or her teacher.

As a student progresses and is ready for more and more personal responsibility the nature and extent of the support gradually changes. Eventually, the time comes for 'the real project'. The student, rather than the teacher, can now provide the detailed structure. Individuals can be expected to choose their own topic, write their own introduction and assignment sheet, plan their own work in detail asking the teacher for advice about resources from time to time, write a report and perhaps even an evaluation. But of course this could not happen without the sort of painstaking teaching, support and confidence-building described above.

In my view fieldwork is really just a 'special study' or a project done with the environment as a context. The amount and type of structure and support can vary in the same way. The environment used can be familiar or unfamiliar. Residential experience is in itself valuable and this can be combined with a disciplined study of a new environment. There are excellent opportunities for cross-curricular work involving art, English, social studies, environmental studies, outdoor pursuits, and so on as well as science. Opportunities are enhanced if a group of teachers get to know an area really well and return several times with different groups of students. If the teaching team covers a range of disciplines then its members too will broaden their horizons by their interactions, including joint involvement in researching the environment.

CASE STUDY 2: PROJECT WORK AND PRACTICAL ASSESSMENT

Mary Doherty

Introduction

How has GCSE physics helped girls? First, there is no longer a need to allocate pupils to 'O' level or CSE groups at the start of the fourth year and, therefore, pupils' confidence is not knocked from the outset. All the girls I now teach understand they have an equal chance to study physics at the level which most suits them. This has been a source of great motivation and despite the fact that our pupils are set in the fourth year into an 'A' group and two 'B' groups, the 'B' groups have insisted (most vociferously) that they should cover the same work as the 'A' group. With one or two exceptions they have done so with success. Under the previous GCE/CSE system there was no need to teach the CSE groups some of the more demanding parts of the 'O' level work and so they were denied the opportunity to show how much they could cope with. I believe that the syllabus encouraged teachers to put false ceilings on pupils' ability.

As in other schools, our experience of practical assessment in the department is very limited and I have been concerned that we should not spend *all* our time assessing practical work. Neither did I want pupils to be unnerved by practical assessment on a particular day. In keeping with the philosophy of GCSE I wanted practical assessment to be part of our normal teaching. After much reflection I thought that an extended practical project would best suit our pupils' needs. It would:

1. Provide an opportunity to explore different teaching approaches.
2. Give pupils the opportunity to set their own pace and allow pupils some say in their learning experience.
3. Give the teacher the best opportunity to assess all pupils during normal lesson time.

For many reasons, I decided that we would have an initial exploratory project which would allow both teachers and pupils an opportunity to come to terms with the demands of practical assessment. I chose to use the Energy Efficiency Office 'Practical Energy Projects'. This consists of 16 projects with a computer program to assist with calculations.

How was the project planned?

Directly after a heavy winter fall of snow and a 3-day school closure due to the bad weather, two classes of pupils were introduced to the project by viewing an 'Energy Factor' video which, despite its tendency to stereotype male and female roles, did contain some very useful ideas. Class discussion followed about the need to conserve energy and how it was wanted. We related the discussion to the needs of the very old and young during a very cold snap. We then introduced the project cards and explained to the pupils what we wanted to happen. Pupils were to choose and develop their own project. We gave the pupils the weekend to plan their projects and to gather any materials they would use. We provided only very basic materials like tin cans, thermometers, and a light meter: the pupils provided all other materials.

The pupils were asked to work in groups where they had common ideas and each group was issued with a tray to store equipment and materials. They were given 5 hours of physics lesson time plus 2 weeks homework time to undertake the project and to record and display their results and conclusions. During this time we teachers marked pupils on observational skills *or* design and planning *or* manipulative skills as appropriate for their project.

In order to avoid what some pupils might perceive as possible teacher bias, we decided that some assessment of 'recording' and 'reporting' skills were to be undertaken by the pupils themselves. We, with each class, discussed the most appropriate methods of doing this, and pupils suggested that each group would give an oral presentation to the class with graphs, results, etc. They were to explain what they had done, what they had found out, what problems they had met, and so on. Each group would have to make clear the contributions of each member of the group, and marks would be awarded for the efforts of individuals. Pupils also felt quite strongly that groups should not be given a time limit for their presentation. The criteria for assessment of recording and reporting were discussed in the class and the pupils decided that if each of them was marked orally and immediately, there would be a tendency for other pupils to be influenced by early markers which they felt would not necessarily be fully objective. Some of the pupils cited an example of a TV programme about drugs where one boy had been led to say he had heard 13 beats of a drum (when in fact he had counted only 12), because six other teenagers had stated that they had heard 13 beats. To overcome this kind of peer pressure, they decided that when the group concluded their presentation, questions could be asked of the presenters and then, on the count of three, all pupils would simultaneously raise in the air a number from 1 to 5 (rather like in ice-skating), which would then be added together and the average found.

What did I learn from the project?

First, some pupils chose projects that I would not have allocated them (probably because of some preconceived notions about their ability to cope) and they all worked with tremendous success. In fact, the motivation of all the pupils exceeded all our expectations. We found that, because all the pupils were fully occupied, it was therefore relatively easy to assess all pupils.

Secondly, I realized that the objectives we generated to be assessed had been too hastily decided upon by me. On reflection, it would have been much more feasible to assess all these areas during the course of the 2 weeks.

Thirdly, the role of technician changed during the assessment projects. They had to be available and to respond to pupils' demands as they modified and adapted experiments.

Fourthly, the work broadened and extended beyond what we might strictly call 'physics'; for example, pupils wrote poems advising the elderly on how to keep warm. Pupils devised, and in the light of experience, modified questionnaires. Some performed little plays to show how the elderly could be informed about insulation and how to best avail themselves of help. One group made cut-out models showing appropriate clothing for summer and winter and also made 3-D models of how the skin behaves when we are hot and cold.

Fifthly, all the pupils found areas in which they could succeed, and in which they could appreciate the successes of others.

The quality and standards of the presentations were in general much better than I had anticipated and showed much ingenuity on the part of pupils. I was disappointed to discover, however, that despite the fact that the quality of work produced by the 'B' group equalled and in some cases exceeded the work of the 'A' group, the 'B' group awarded themselves 3's or 4's whereas the 'A' group awarded themselves 4's or 5's.

Interestingly, the most able pupils stuck strictly to the letter of their project, which were often the most mathematical. They undertook it in the most 'perfect' manner, but it was well within their capabilities, and unlike *all* other groups did not extend their project in any way. When I questioned them about this, they said they had always been told that they should answer the question and that there would be no extra marks for extra effort.

What did pupils learn from the project?

First, they learnt a little of what a GCSE practical assessment means. And, happily, most pupils discovered they were much more capable than they had realized. Others discovered how easy it was to go wrong and how frustrating this can be. All pupils learnt new skills. For example, they learnt communication skills, planning and pacing, working to deadlines, competitive skill, and working with the minimum of supervision. All the pupils stated that they had enjoyed the work and wanted to know when the next project would be and how many more they would have to do. They did feel that in future they would like the project to be less prescriptive and more open-ended.

I intend to plan three further extended projects in which to conduct practical assessment and I will bear the following points in mind:

1. To avoid projects which are too prescriptive – and not to underestimate the capabilities of pupils.
2. To check very carefully the wording of the instructions. Pupils should not be penalized because the teacher did not plan the project adequately.
3. If possible, to trial the project with one group and assess with a different group. This allows opportunities to iron out wrinkles.
4. To change our system of allocating pupils to GCSE groups. In the light of this experience, the 'B' group did not live up to its label.
5. To realize that extended projects are demanding and tiring, and so to check other subject demands and to time the projects accordingly.

The projects involved pupils working both in the laboratory and on the school campus. Fortunately, colleagues were very cooperative and did not seem to mind in the least the groups of students recording the amount of light in various parts of their rooms, taking temperatures, or setting up test double-glazing units. The pupils were well behaved and could be trusted to move round the school in a responsible manner and to cause the minimum inconvenience to others. Although the Science Department is housed in a new purpose-built suite of four laboratories which are very well equipped, the project could be carried out in most school labs. Because the pupils provided most of the materials, the project's success relied much more on the use of the pupil as a resource than on a wealth of expensive school-based resources.

CASE STUDY 3: PRACTICAL ACTIVITY – WHOSE PROBLEM IS IT?

Philip Naylor

Practical work is fine if you intend to develop particular skills such as recording, measuring or drawing conclusions, but what else do our pupils gain from such experiences? We enter the classroom with a grand educational master plan (without telling the pupils) and expect them to 'see' what we are about, picking up whatever clues we may care to offer and hopefully as time passes become aware of the meaning of *science*. Teaching must be and is a lot more than that. If, for example, we approach the class with a 'problem-solving' situation, the first question that is asked by all pupils is 'Is it my problem?' and, unfortunately, many pupils do *not* consider it to be their problem and coast through the experience without touching them.

I have become increasingly committed to practical work as a means of pupils developing skills but, more importantly, it is a purposeful way of learning about the world. Although this approach may seem to be time-consuming it is only at the expense of my perceptions of how to 'get' pupils through examinations. What pupils actually learn about the world in which they live from many school experiences can possibly be summed up by comments heard from my colleagues that they 'learned about the real world despite their education'.

The group of 21 pupils I worked with were studying aspects of hygiene in the context of a Food Technology module. The learning outcomes that were part of the course included knowledge of the methods of preserving food, conditions which cause food to 'go off' and how to produce a bacterial culture. In order to allow the study to become the pupils' investigation I posed questions that related loosely to the syllabus content (which had already been discussed with pupils) and allowed groups to consider as a result of their own experiences what they could do to find out more. Small group work activities were organized for the class to help them discover how they might find out about these things. This led to discussion about the new 'McDonalds' fast food store that had opened in the town and various comments on 'how clean' they seemed to be. The pupils had begun to make the learning activity real and purposeful by placing it in a context that was important and relevant to them. This also had important relevance post-school, since it was a possible form of employment in the not too distant future.

Clearly, encouraging pupils to 'own' the activity results in a wide variety of investigations being suggested, and when attempting to set up these often diverse activities the support of a technician is important. Perhaps more important is the realization that the pupils are capable of working with real-life constraints, without full resources, yet they manage to produce workable situations. Such a 'problem-solving' approach to a real-life difficulty is something that pupils have to develop out of school, so why not use it – provided you accept the likelihood of non-standard practical situations being set up on the odd occasion.

As a result of their small group work and subsequent discussion, the pupils decided that their study should be based on a trip to McDonalds to investigate the reasons for the apparent cleanliness. In reality, many wanted to visit to find out about possible jobs and to see if they could get free samples! It is important to accept the pupils' contributions to such an activity, especially as such contributions provide honest motivation, purposeful activity and, in this instance, brought to the fore the relationship between cleanliness and conditions for bacterial growth.

The class divided the study into various parts, which were taken on by different groups comprised of three or four pupils. One of the agreed objectives of each group was that it would report its findings to the whole class. Some groups took on the problem of finding out how the food was stored, because they all understood the need for food stocks to be kept;

some looked at the way in which the staff were treated and what their working conditions were like (this group had some ideas about future employment); and some considered the need for tests that could indicate just how clean the place was. This latter group spent some time considering the need for samples and how they could culture such samples to produce a fair test. My task as a manager and facilitator required a little tact, especially when the idea of sampling was suggested! I was extremely heartened by the support and subsequent treatment we all received from the staff at McDonalds and would recommend this visit to others. (The free samples were very tasty!)

Classroom activities during and after the visits consisted mainly of discussions about specific scientific techniques or ways around a particular problem, and it was here that my expertise as a teacher was most valuable. Very soon the pupils became increasingly aware of the need to report back something specific to the class as a whole and with a little prompting set their own targets (criteria for success) on their work. It is important that they do set their *own* parameters to *their* activities and see the importance of evaluating the effectiveness of their own work if they are to see the value of assessment strategies.

The flavour and direction of discussions were wide-ranging, covering many issues of scientific investigation. Different theories, hypotheses and ideas were put forward by group members and debated for their potential value in helping to solve the problem. For example, one group, looking at 'cleanliness', had put forward the idea (hypothesized) that bacteria were present in the air and that was one of the reasons that the staff were constantly wiping down all the surfaces. They were also aware of the fact that there was a wash basin in continual use, and they theorized that perhaps people working there brought 'their own' bacteria and thus workers' hands should be sampled. Another idea considered by the group was that such meticulous cleaning was a company sales gimmick since group members – and probably the workers – never really went to this trouble at home. After some heated debate the group eventually settled on the first possible explanation. They decided to take samples from surfaces and from the air using sterilized cotton buds and developed a rather smooth technique of opening the dish quickly, smearing with a selected sample and sealing it with prepared adhesive tape. The reason for their particular style became an integral part of their 'reporting back', and laid the ground for discussion about fair tests and the existence of aerial bacteria. The group that looked at working conditions had devised a rather basic questionnaire but had specific ideas from their own experiences about what would constitute a 'good' working environment.

Reporting back provided a situation for consolidation, especially as the pupils were all aware of the work the other groups had completed. The consolidation process was aided by some questions that I had prepared in relation to the syllabus content, but answers were forthcoming in terms of the real-life situation they had visited and supported by additional experiences of their own from previous lessons or from home. The discussion following the reporting back was focused around two issues. One was based on the role taken by some of the trainee managers and the way in which girls did not seem to have the same opportunities for a variety of occupational roles. The discussion also pinpointed a general split between the roles perceived by the male and female pupils which to some of the class was unacceptable. The second issue was as a criticism of the 'bacterial sampling' group. It appeared from their results that the sample taken from the anti-bacterial soap was the most contaminated. This raised questions from their peers as to their sampling methods and experimental techniques. Subsequent discussion then ranged on the problem of a fair test and whether or not the techniques that had been used were good enough.

Early on in the process of establishing what tasks each group was to do, the pupils had established criteria by which they could assess the worth of their activities. They had produced some of their own assessment criteria in conjunction with those that I felt appropriate. The sharing of such assessment criteria was important in that the pupils could estimate the relative success of their ventures and they could also appreciate the reasons behind the assessment strategies that I was using.

The total activity took three double periods (1½ weeks) with the use of some time from the personal and social education teacher and the Head of Upper School; the learning outcomes from the syllabus had been achieved. Practical skills had clearly been developed but the pupils had been involved in a lot more. Such things as the ability to work with others, the demands of team work that gave them *all* responsibilities, and the realization that all people are not treated equally were all aspects of the learning situation that might not have occurred in the context of a different approach. The approach was chosen deliberately to allow a much wider learning situation to develop. This is with the caveat that real learning is proportional to the value of what is being taught in relation to the real world of the pupils in the class. Despite the fact that such approaches are considered more time-consuming, the amount of curriculum time was not much greater than that of a more traditional approach, and the involvement, challenge and enthusiasm of the pupils was significantly greater.

CASE STUDY 4: SCIENCE PROCESS INVESTIGATIONS

Di Bentley

Practical experiences in science are commonplace. The justification for including such experiences in the science curriculum has been that science is an empirical subject and young people need to experience the processes of science in order to understand them. In this way, investigations which help youngsters to explore how hypotheses might be set up, variables controlled and the relationships between them examined are the 'bread and butter' of many science lessons.

In many ways, the two lessons described here are no different to hundreds of lessons which take place in science laboratories with first years every day. They began with a hypothesis, and used a process of scientific investigation to explore the answer to the hypothesis. The lessons took place in a mixed-ability class of 27 first years who had come from a variety of different primary schools just one term previously. Some of the primary schools had science in their curriculum, others had not, so the youngsters had a variety of scientific experience in formal learning terms to bring to the secondary school lessons. The science course the youngsters had been following had focused particularly on raising pupils' awareness to the skills they would need to use science as an investigatory process. The lessons described here come from the unit on decision-making and analysing skills. The course started at the beginning of the year by referring to the pupils as researchers and myself, the teacher, as the 'senior researcher'. As senior researcher, I was in charge of budgets and required to check all investigation plans produced by the 'researchers' for their cost-effectiveness and feasibility.

In the first lesson, on entry to the room, pupils were each given a card which contained information on the case study they were to use and details of their working group. Working group allocation was by means of a number and a letter. The pupils were asked to find those people whose card had the same number as theirs. This effectively divided the class into groups of four or five. The groups appeared to be random, but were in fact carefully organized. The youngsters had been told at the beginning of the course that during their first year they would have the opportunity to work with everyone in the class, in order to learn as much from other people as possible. In actuality there was another hidden agenda to this. It enabled the construction of groups of different gender and racial mix on various occasions, as well as ensuring that all youngsters got the opportunity to monitor their skills of communication and cooperation in as many different circumstances as possible.

The groups collected their brief case studies. The case studies for each group were different, but directed the youngsters to similar problems being experienced by particular pupils of their own age. They were asked after reading the case study, to do the following things:

1. Brainstorm for 2 minutes the possible decisions the pupil might make to solve the problem.
2. Discuss and write down what extra information they would need to help them to solve the problem.
3. Plan an investigation that would provide some of this information.
4. Write a list of apparatus, describe the plans for their investigation to be carried out next lesson and submit them to the 'senior researcher'.
5. Select someone to report their findings and recommendations at the end of the investigation.

Preparation of the investigation plans often required a great deal of teacher input. Many youngsters, despite several weeks of working in this way, were still very unsure of themselves and would rather be told what to do. Thus a set of 'hint sheets' were kept available to lead these youngsters into their work in a more structured way. After they had completed the tasks, they were asked to do their background research into the problem by reading around various aspects on which they had focused during their discussions. This aspect of the lesson took a great deal of prior preparation, ensuring that books of the right level were available, relevant magazine articles collected, and that the school librarian was primed for the sorts of requests that were to come her way. Sometimes it meant producing digests of research information that were written in terms too difficult for first years to understand.

The pupils' investigation plans were handed in at the end of the lesson. In the interval between this lesson and the next, the task for me was to collate the lists of apparatus and make changes where appropriate, e.g. when apparatus was unavailable or simply not feasible. I recorded all these changes or alternative suggestions for the pupils on a 'Senior Researcher's Report Sheet' which I attached to the front of their plans. The apparatus was ordered from the technicians and collected in boxes marked group 1, 2, etc. so that the pupils would not have to search for bits of apparatus among a general collection.

At the beginning of the second lesson, the pupils collected their research plans and the 'Senior Researcher's Report' as soon as they came in. Some wanted to discuss and question any changes in their plans with me, whereas others simply wanted to get on with the investigation. So the start of the lesson was a staggered one for many pupils. The 'Senior Researcher's Report' gave them a strict time allowance for the investigation – they had to be ready to report back by the end of the first lesson (40 minutes). During that time, it instructed them that, as usual, they also had to assess their working colleagues for their abilities to cooperate, communicate their ideas to the group and listen to others.

Inevitably, because the case studies were all different, the reporting back of each group differed a great deal. However, the case studies had been written to focus attention on one particular aspect which could be handled by a scientific investigation, so the experimentation plans that the pupils prepared, while taking different approaches, all focused around the same idea. Perhaps the easiest way of giving a flavour of the lesson is to examine the report of one particular group.

This group had a case study about a shy girl who was the only one in a family of brothers. Her problem was that her mum was elderly, past menopausal age and was a poor source of advice about any issues of puberty. In her primary school, the girl had been given information about menstruation but had not yet started menstruating. She was not sure what type of sanitary towels might suit her best but, being shy, the one thing that was very important to her was that the sanitary towels should be efficient in their absorbency. She had looked at them in the chemist shop, but was unsure what the difference was between 'normal', 'super' and 'super-plus' apart from the fact that 'super-plus' ones were more expensive.

The group identified several decisions facing the girl. The most important was the matter of absorbency and the second was that of the type of sanitary protection, internal or external. There were several other decisions she faced as well and they mentioned these in their report, but felt that as scientists they could only really investigate the first two. They planned an experiment which involved different grades of sanitary protection. They requested small, same-size samples of different grades of internal and external sanitary protection from a variety of manufacturers, a dropper, water and a viscous fluid such as glycerine (they argued that the consistency of the menstrual flow might not be the same as water). In their investigation, they tried dropping water from different heights on to the sample and timing how long it took for it to allow the fluid to soak through. They reported having hit a snag, however, when using internal sanitary protection in this way, in that some makes were more densely packed than others and so their results were not easily comparable. Unpacking the sample was not a true reflection of how it operated in reality, and they declared that they had not been able to solve this problem in the time available. Nor had their glycerine produced conclusive results in comparison with water and they felt they needed extra time for this too.

Overall, all the groups took some 30 minutes to report back and a variety of advice on absorbency was provided for the people in the different case studies. Several youngsters felt that unsolved problems were ones they might follow up in their own time, and mentioned that they would ask older sisters and mums what absorbency problems they had experienced One group (an all-girls group who were usually fairly inhibited in science) offered to compile a group report on the kinds of problems facing different women of different ages by conducting an investigation at home. Another group wanted to learn more about the 'fluid consistency' problem, and asked to come to science club to investigate the absorbency of fluids of different consistencies. After further discussions along these lines, I invited the pupils to spend their next 2 weeks homework time following up any aspects of the problem that interested them and present a written homework report. They were reminded to complete their personal skills assessment sheets for the lesson and to complete one for their homework reports as well, saying which skills they thought they had used and how these had been developed by the work.

What did the pupils get out of it? For the girls, the context of the problem meant that real issues which they faced as important ones in their lives were given validity as aspects for scientific investigation. The continual reminders of the skills-based aspects of the investigations made plain to pupils the need to monitor and improve their own performance. Their treatment as responsible 'researchers' capable of planning their own work as scientists raised their awareness as to what science might be about over and above a 'body of knowledge', in terms of the way scientists operate and the processes involved in science. The approach also capitalized on the degree of responsibility that pupils brought from their primary school experiences, stimulated their interest and did much to remove their embarrassment in discussing issues of puberty. They requested an 'information briefing' about puberty and menstruation to answer some of the questions that the literature sources I had been able to supply left hanging in the air.

SUMMARY

In this chapter we have focused on practical work and projects. As a summary, we consider four questions about them as techniques:

1. Which aspects of active learning do they encourage?
2. What planning do they require?
3. What organization is needed?
4. What else is important?

1. Active Learning

The youngsters in our first three cases show several of the characteristics of active learners. For example, they display the ability to initiate and take responsibility for their own work. They seem to feel in control of their own learning – to such an extent that they are willing to make decisions and solve problems. Project work is a step towards encouraging autonomous learning.

Mary Doherty and Philip Naylor describe pupils who show a high degree of organization: they are well able to pace their work and meet deadlines. They also engage in peer evaluation.

2. Planning

All practical work requires planning. The case studies show that initial planning is mostly about collecting potentially useful apparatus, written material and computer-based resources. Inevitably, the 'fine tuning' of resources can only take place after pupils have identified their project areas. In very open-ended project work, almost anything could be needed. In case study 2, Mary Doherty solved this problem by saying that the school would supply only very basic equipment, the pupils themselves having to supply anything extra. This is an interesting way of helping pupils realize that resources are finite, and that their plans must be made within certain constraints. Mary Doherty devised two constraints: the schools' supplies and pupils' own ingenuity. In contrast, Di Bentley did not ask youngsters to provide their own resources but acted as a resource manager by playing the part of 'senior researcher'. Her task was to control the 'budgets'.

Philip Naylor includes other aspects of planning, such as the organization of the visit to the restaurant. This may require preliminary visits by the teacher (a 'Comprehensive Good Food Guide'?) – the personnel at the receiving-end need careful preparation too. For instance, they need to know numbers, time and duration of visit, have some idea of the purpose of the work, the kinds of questions youngsters might ask, the requests they might make, and have some awareness of the levels of communication required. The timing of such planning is crucial if the visit is to be successful and mutual respect between school and community is to be maintained.

3. Organization

The organization of time and space are important, particularly with project work. Often youngsters need to be able to store work from one session to the next and to have ready access to it. Youngsters may want to work on some aspects of the project outside lesson times. The class teacher needs to be flexible. Colleagues may need to be warned that some youngsters will be around the science department at unusual times. Teachers elsewhere in the school, as in case study 2, may need to be warned that, during science lessons, some youngsters may come into their lessons, or be about the school and/or community. Technicians will need to be closely involved in the planning and organization. They need to know what is to be used in advance, which parts of the equipment are not required after each lesson, and which are needed again. Trays, boxes, labels and storage areas (as well as a 'discards box') are all important organizational features of projects and prolonged practical work. Apparatus may need to be ordered directly by youngsters and it is important, if lessons are to run smoothly, that there is an order system that is understood by all. Pupils, like teachers, may have to come to terms with the fact that if they do not plan in time they will not receive what they require, and will have to reorganize their activities for the session.

4. Other points

All four case studies draw attention to different methods of reporting practical and project work. In the second case study, the pupils chose to have verbal report-back sessions which were assessed by their peers. The pupils in case study 3 also chose verbal report-back, though in their case the assessment was less formal. Visual reporting, too, is a useful means of communicating results. This also provides opportunities for youngsters who lack confidence in talking or writing to participate by drawing, performing or cartooning. Displays of materials can provide a means of communicating the work of youngsters to peers, parents, the rest of the school and the community. Some schools provide visual displays for local education offices, building societies, banks or community halls. Such displays do not need to be the 'best of the art department'. Some schools have gone further by having older pupils photograph and mount the work of younger ones as part of their photography work in science. This is a particularly valuable way of recording reports which are in the form of mime or drama. Note, too, that project work comes with built-in homework.

Finally, then, what do youngsters gain from such approaches to teaching and learning? We think Mary Doherty sums it up well when she says:

> pupils learnt new skills. For example, they learnt communication skills; planning and pacing; working to deadlines; competitive skill and working with the minimum of supervision. They all stated that they had enjoyed the work . . . and felt that in future they would like the project to be more open-ended.

BIBLIOGRAPHY

Beatty, A. and Woolnough, B. (1982). Why do practical work in 11–13 science? *School Science Review* **63** (225), 768–70.

Department of Education and Science (1985). *Science Education 5–16: A Statement of Policy*. London: HMSO.

Her Majesty's Inspectorate (1987). *Report by HM Inspectors on Survey of Science in Years 1–3 of some Secondary Schools in Greenwich*. London: DES.

3: TALKING AND WRITING FOR LEARNING

INTRODUCTION

This chapter looks at two aspects of language in science. Both classroom talk, particularly group discussion, and effective writing allow youngsters to formulate, reformulate and quite literally 'come to terms' with their understandings of scientific ideas. Talk can take many forms, from informal conversation during an experimental session to structured, whole-class discussion. It is also very often taken for granted. For example, in many published schemes in science, experimental work is described in a wealth of detail. 'Discussion lessons', on the other hand, contain little guidance to teachers on how to conduct such a session. The feeling is that talk happens naturally and there is no need to plan for how it should take place.

Similarly, criticisms of science teaching have often centred on the abuse (or uncritical use) of class worksheets which make few demands on youngsters' writing skills. Reports on experiments are usually quite cursory and it is still quite common to see laboratory reports written under the headings 'apparatus', 'diagram', 'method', 'results', 'conclusions', etc., all in the traditional third-person-passive style. Our contributors in this chapter describe how they have used both classroom talk and writing to help develop and explore ideas in science.

We look first at classroom talk. As a result of curriculum development work in the late 1960s, the myth has developed that all science lessons must be practically based. Nowadays, there are more science lessons which feature discussion as a means of introducing or helping youngsters to understand aspects of, say, the social and moral issues of science. Whatever the activities designed for pupils during a lesson, whether practical or theoretical, we believe they will need at some time to talk about their thoughts and ideas with peers.

What does it mean to urge more discussion in science and, to be more precise, what *is* discussion? There are many different types of talk, e.g. teacher–pupil talk and pupil–pupil talk. Some, designed to draw out specific points and answer questions which seek information, might be tightly controlled. Some can be more informal and intended, say, to pool results of an experiment. Science lessons, however, are dominated by question-and-answer sessions. They are often referred to as class discussions but usually consist of a recap of the last lesson, or the introduction of new ideas that takes place around the teacher's bench at the start of a new piece of work. The extract below is an example of this type of 'whole class discussion':

Teacher:	The earth spins about an axis. The point at the top about which it spins is called what? What's it called?
Pupil 1:	The geographical axis.
Teacher:	Geographical . . . ? The point at the top. Almost right. Yes David?
Pupil 2:	North.
Teacher:	Yes, north. And the point at which the magnet appears to be – at the top there, just underneath the ground – what's the name of that point there?
Pupil 3:	Magnetic north.
Teacher:	Magnetic north. If I was standing on the top of the magnetic north with a compass, what would it do?
Pupil 4:	It would go round and round and round.

Here the purpose of talk is to elicit information, not to explore and investigate youngsters' own ideas. The teacher asks very closed questions and, in doing so, shapes everything the youngsters say. Can we then refer to such prescribed talk as a discussion?

Button (1974) distinguishes between two types of discussion. He describes 'horizontal discussion' which ranges over the surface and serves to alert the pupils to the width of topics, test knowledge and comprehension. 'Vertical discussion', which is more limited in its scope, enables probing for deeper meaning. The type of discussion in the extract above would fit best with Button's horizontal discussion. It serves a useful purpose, but should not be the pupils' only experience of discussion in science lessons. Active learning would seem to require periods of vertical discussion to enable the sharing of ideas and theories about science and scientific phenomena.

What would we need to do to ensure that pupils have access to good vertical discussions in science classrooms? McClelland (1983) claims there are four necessary conditions for 'genuine' discussion to occur:

1. The topic needs to be problematic, but within the scope of the knowledge of the discussants.

2. Ideally, the discussion should be between 'consenting peers'. The involvement of a teacher in class discussion should be under circumstances where their status does not hinder the discussion, where deference is not automatically made to their view.
3. Group size must not be so large that it allows separation into active and passive contributors.
4. Discussants need to feel that a worthwhile outcome is possible, and be sufficiently interested to engage with the task.

To encourage active learning through vertical discussion, then, we need to meet each of these conditions – McClelland claims that anything else is 'pseudo-discussion'. His analysis, however, examines only the conditions that need to be present. It tells us little of the processes we might wish youngsters to experience. The Association for Science Education (ASE, 1981) suggested three characteristics of the process of discussion:

- It begins from common experiences.
- The participants pool ideas, without at first necessarily judging the relative merits of the ideas.
- The discussion involves evaluating, rejecting and selecting ideas from among those being offered. This process should be within the control of the group of participants.

We would add two further points:

- Discussion should result in some change of ideas by the participants. They may well need support to help them accomplish this. In actuality it means working in an atmosphere in which they can trust their fellow participants – and the teacher – not to dismiss their responses or subject them to ridicule.
- Discussion should only terminate with feelings of satisfaction for all when a common solution or idea has been agreed upon. That is, feelings of rancour, anger and irritation sometimes aroused during the discussion need to be dealt with. It is the teacher's responsibility to be aware of the mood of discussions and create the conditions for ill-feeling to dissipate.

To organize good (vertical) discussion, then, teachers need to both set prior conditions and help youngsters understand the process. Many of these issues are featured in Mike Watts' contribution, 'Discussing Physics'. He spends time focusing the youngsters' attention on the processes of discussion, before the second part of the lesson about gravity. In other words, the prior conditions and the necessary skills and processes of discussion become points to be formally taught in the lesson. Harry Moore's study also features overt teaching of discussion skills. He focuses on listening skills, where youngsters are taught to be aware of their own skills in listening as one part of the process of discussion. Pauline Hoyle's contribution describes pupil–pupil talk and outlines their tentative sharing of ideas, their suggestions for how to proceed (sorting out the agenda), and their general language development. In particular, she focuses on the power talk gives to pupils whose first language is not English – the power to actively explore concepts and gain help from their peers.

Clive Carré (1981) illustrates several forms of classroom writing in science. He makes the point that much of the stylized reporting of scientific experiments which pupils are required to do is somewhat useless, because they have not actually understood the science at issue. He suggests that writing in science could be more purposeful for both teachers and pupils. It might, for instance, encourage youngsters to 'reflect upon their experiences in the laboratory and in their world'. Writing can help teachers to help pupils to relate to what is new, and to what is already known – primarily, he argues, by choosing alternative writing tasks which invite a personal approach, using youngsters' own words to explore ideas about things they know.

There are two contributions which show what such alternative writing tasks might be like. Steve Whitworth explores how extended discussion, linked to writing, boosts the confidence and self-esteem of low-achieving youngsters. At the same time it allows them to re-vise (revisualize) and experiment with ideas in science for another audience. Mike Watts' second case study encourages youngsters to use poetry to express ideas – and feelings – about what they have learnt. As with Steve Whitworth's pupils, the youngsters discuss the issues before committing them to writing. What makes this writing different is that it is in a form not common in science, i.e. poetry.

The final contribution, from Brigid Bubel, is a very different one. It is the only contribution in the book which addresses the issue of assessment for itself. With the advent of the GCSE and its emphasis on course-work, many teachers have found that their teaching approaches in the fourth and fifth years have had to change. Brigid's work was done just prior to GCSE, but is none the less still relevant as a means of assessment. She makes the point that pupils' writing about their work – such as design of experiments, describing the point and purpose of the social implications of science – often does not do justice to their understanding. There is evidence from the APU (1982) that when students are invited to talk about how they would set up and conduct an experiment, their performance is higher than when they are asked to write about the same setting up and conducting. It seems possible that in writing about their plans, they take a great deal for granted; i.e. because they know how to perform a particular process well, they do not bother to write down exactly what they would do. Talking about the process instead seems to encourage them to be fully descriptive. Therefore, many examination situations may not really help students to show what they 'know and can do', if the only medium for that showing is writing.

Brigid describes how the skill of communication was assessed by using an oral format, and how this helped pupils to be more explicit about their understandings of science. Since many GCSE science examinations also assess communication skills, this may be an aspect that readers wish to explore further.

CASE STUDY 5: DISCUSSING PHYSICS

Mike Watts

Introduction

You don't have to be an educational guru to appreciate the value of group discussion work in science. That said, it is something that has been promoted time and again by many educational gurus – without, it seems to me, too much effect generally on science education. Most science teachers would say that they *do* use discussion work in class. What they often have in mind is the usual question-and-answer session that starts a lesson, say, on pressure. Or they think of the small groups talking in a huddle round a clump of apparatus, trying to sort out what is supposed to be happening. What seldom happens is that teachers fully plan discussions, prepare for them and, perhaps most importantly, teach youngsters *how* to manage discussion sessions. There is often an impression that since youngsters are, on the whole, good at talking and can often talk to the exclusion of the other things they are supposed to be doing, there is no need to train them in the art. For the cynic, teaching kids discussion techniques smacks of masochism.

This section draws on two consecutive lessons in a week, the first a single, the second a double. The class (28 fourth year, mixed 'O' level and CSE physics students, with a 3:1 ratio of boys to girls) were finishing off work on energy and forces, and the teacher wanted to focus on gravity. I had been working alongside the class as part of a research programme and joined in fully with the teacher for these sessions. Over the term the class had, periodically, been answering some physics questions for my research and in this instance we wanted to use the questions to better benefit. The youngsters had been experiencing some considerable conceptual difficulties with ideas like mass, weight, pressure, force and gravity. Our plan was as follows. In the single lesson we would spend time teasing out some of the ways in which discussion groups might work and what the outcomes might be. We would then set the class some discussion tasks to finish off the 40 minutes. In the second lesson (80 minutes) we would return to some of the issues raised, push for further discussion, have a plenary session where we 'summed up', follow this with some practical work, then tidy up, set homework, and so on. The questions themselves were based on a series of pictures and the results of the research have been discussed elsewhere (Watts, 1982; Gilbert and Watts, 1983; Watts and Gilbert, 1984). They are shown in Figs 5.1 and 5.2: Fig. 5.1 shows a series of simple line drawings; Fig 5.2, a set of 'multiple-choice plus explanations' questions.

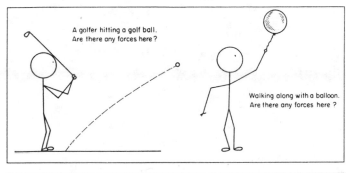

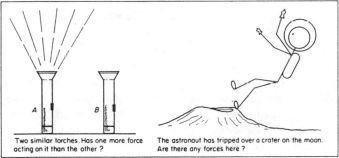

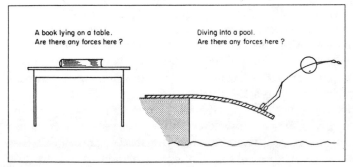

Figure 5.1 Some examples of the cards

Figure 5.2   An example of the multiple choice questions

Here is an astronaut on the moon.
He has gently let go of his spanner.

Which drawing do you think best shows the GRAVITY on the spanner?

NONE

The first lesson: part 1

We began by explaining the structure of the lesson to the class, that we were going to 'discuss discussion' and then talk about some physics questions. The pupils were first asked to work in pairs and to make a list of what discussion was like in different lessons – science, English, mathematics, and so on. They were told to work quickly and that they had only 3 minutes. We kept them fairly much to time because we had a very tight schedule. Then, still in pairs, they had 5 minutes to answer two questions: what is the point of discussion work and what makes a good discussion?

From this they had to share their ideas. They were asked to move into bigger groups (two groups of six and two groups of eight) where they had to pool their thoughts on the last two questions, appoint a spokesperson, and be prepared to report back to the whole group. Again they had to work quickly. After they had composed their final lists, we heard all four reports and collated them on an overhead projector (OHP) at the front of the class, all of which took about 15 minutes.

Their responses were both varied and predictable. Discussions in other lessons were sometimes stilted question-and-answer sessions ('you just have to think up the right answer to their questions'), mock debates, discussion huddles ('we talked about last night's television') or even, in an English lesson, 'talking to a tape recorder'.

They thought the purpose of discussion was to share ideas, learn from each other, appreciate other peoples' opinions, and to avoid doing any writing. They thought a good discussion was where everyone got a chance to talk without being 'cut out', though some thought it was when 'you have a good argument'.

We then spent a few minutes adding our own items to the list. We thought there needed to be:

- a structure (a begining and an end);
- something tangible on which to focus;
- some specific outcome from a discussion session;
- no particular reason for it to result in consensus – there was value in clarifying divided opinions;
- a clear opportunity for people to listen carefully to what was being said.

We rather stressed the notion of everyone having a chance to talk and the need to listen. And then said we would leave the list up on the screen while we moved into the next part of the lesson.

The first lesson: part 2

The task was to discuss the questions about gravity. The pupils were asked to make groups of four from the original pairs and to tackle the questions in Fig. 5.1 for the 10 minutes left of the lesson. They were to answer the question about force or gravity on the picture and, in the back of their books, write a sentence of explanation. We would consider their responses in the next lesson.

The second lesson: part 1

The pupils were asked to form the same groups of four as at the end of the last lesson and, for 5 minutes, revisit and share their answers to the questions. They were then given the multiple-choice questions to answer in 10 minutes. Still in the groups, they were to tick the option(s) they thought were appropriate and write a sentence of explanation.

The second lesson: part 2

The plenary session that followed was quite long and sometimes difficult to keep in check. The teacher fielded the responses while I collated them on the OHP as a list of questions to be answered. We began by asking one member of each group to report on how well their discussions had progressed. They all thought the 'guidelines' had allowed 'good dis-cussion' of gravity. People had actually taken turns to talk, and some admitted even having listened. We then moved into a whole-class discussion.

There was clearly a range of different interpretations about what gravity is, how it works, what effects it has, how it differs from place to place, and so on. Some of these we were prepared for, because they matched what had surfaced in other lessons and in other research studies. The greatest sticking points were whether gravity operated on things as they were moving or thrown upwards, and how it worked on the moon. The class was fairly divided on both issues, though there were shades of opinion within camps. One common perspective, for example, was that gravity on the moon is, in comparison to that of the earth, both reduced and reversed. That is, it is weaker than the earth's and works upwards – an astronaut's spanner would move (float?) upwards when released. It is only the astronaut's heavy boots that keep him tethered to the moon's surface. This was so, apparently, because there was no atmosphere on the moon. The debate in the class ranged far and wide, lasted nearly 40 minutes, was marvellously enjoyable, often very funny, and clearly highly stimulating. We gauged the time to draw it to a close when more questions were raised than could be answered, and when it was beginning to become focused on some of the more vocal participants to the exclusion of others.

The second lesson: part 3

There was no immediate way to attempt to resolve all the conceptual problems that were made apparent, though we did say we would keep returning to the OHP list in the following lessons. The teacher had organized to do one practical experiment (the 'guinea and feather' experiment) where objects of different mass are allowed to fall in an evacuated tube. The purpose is to show that, in the absence of air, objects of very different mass (and therefore weight) will fall (and not rise!) at the same rate and reach the bottom of the tube at the same instant. Due to a shortage of time, this turned from being a group practical into a demonstration. At this point the discussion time was officially over and the end of the lesson followed the more traditional pattern of writing details, recording observations, setting homework, and so on. For many, though, the demonstration singularly failed to settle the issue and discussions were continuing even as they wrote, packed their bags and walked down the corridor.

Summary

There are very many occasions when it is appropriate to spend class time in discussion. These may be, for example, about the personal or social implications of science and technology, some of the limitations of science as a way of solving problems or, as in this case, about personal and orthodox ways of conceptualizing parts of physics. Discussion can be highly motivating, a profitable way of using time and, if planned, can lead to a variety of important curricular outcomes. Most importantly, it helps to give youngsters the opportunity to explore and share their own ideas, to validate their stake in, and ownership of, science lessons.

CASE STUDY 6: ENHANCING LISTENING SKILLS IN SCIENCE

Harry Moore

The South Devon working group of the Secondary Science Curriculum Review (1986) had as its aim 'to create a bank of flexible resources for teachers to enhance their teaching of science skills and to suggest a series of alternative strategies to the teaching of these within the school curriculum'. Listening is an important stage in learning and a basic communication skill which normally receives little attention in science. What do we mean by skill in this context? The definition of a skill which I prefer is 'a skill is a specific activity which a student can be trained to do' (SSCR, 1984). As science teachers we have excellent opportunities to develop listening, which is a basic skill. Pupils should be encouraged to improve the skill of listening and identify weaknesses, e.g. in recalling facts, reorganizing important ideas and following instructions to perform tasks. The group devised a wide-ranging collection of exercises designed to encourage pupils to be aware of how well they listened and improve their already existing skills. These activities, consisting of both short 'lesson starters' and longer pieces, were tested extensively over a trial period of 1 year with a mixed-ability group of 29 first-year pupils in an 11–18 comprehensive school. Some activities have been used as a stimulus to start a lesson, others in form tutorials and one was used for the whole period.

If an activity has as its aim to enhance pupils' listening skills then we need to set up a system which attempts to identify whether or not this aim has been achieved. Clearly, in a class of 29, no teacher can observe all of the pupils, so I decided to select a group of 6 pupils for observation at different times during the year. The highly motivated and attentive pupils were ignored (there are still a number of these in each group!) and the extreme case of a Special Educational Needs (SEN) pupil. The selected group represented a good cross-section of ability and social background. Observations were sometimes made with the help of colleagues but most were my subjective judgements and records.

I do not claim any significant improvement to have been made but the activities have provided a stimulus to lessons and certainly focuses attention on the skill of listening. The three examples given below show the range of activities tried. They are included in the hope that they might encourage others to try to adopt different strategies in their teaching.

1. Child's play

This exercise, based on the well-known ITV programme, is used to identify items of equipment and to familiarize pupils with the laboratory. All the equipment is out of sight and the group is divided into small teams. As the game is played, all of the equipment required for an investigation appears on the tables. It has the merit of making savings on technician and teacher time in setting out equipment as well as providing an enjoyable stimulus to listening, because clues are given only once. My group found this 10-minute activity a refreshing change.

A further use came in the period before their end-of-unit assessment. Definitions of words and equipment were read out by individuals in order to describe their meaning to others who then were asked to identify the word. This reinforces the requirement for precision in making oral statements (including my own!). Once again pupils must concentrate on listening carefully.

2. Demonstration

Many of us seem reluctant to perform demonstrations these days because of the demand for investigative work and individualized learning. A demonstration of some practical technique to be used in the lesson needs careful observation and listening. The preparation of carbon dioxide and its detection with lime water was demonstrated with pupils around the bench. My usual procedure is to question a few individuals to determine that they have followed the technique, on the assumption that the remainder will also hopefully perform it. However, in this case, as an alternative, pupils were placed in small groups prior to the demonstration and one person was designated to relate the procedure they have all just seen to the remainder of the group. When a group reached agreement on what to do I was called to that table to listen and, if satisfied, allow them to proceed. This may seem a rather time-consuming exercise but, used occasionally, it emphasizes the importance of observing and listening carefully to both teacher and fellow pupils.

3. This is what we have to do

This activity is simply an alternative to that of a worksheet, providing formal instructions for an experiment. A set of instructions is cut up and distributed at random around each small group. Every member of a group must read out their instruction to the others in turn. Agreement is reached on the order in which the instructions should be placed to carry out the experiment. The group then perform the experiment as per their procedure. Some groups will find their procedure produces a different result to others and this provides a useful discussion area itself. If successful, this exercise will encourage pupils to listen to each other and to organize their work logically.

Evaluating the process

Activities such as the above do provide a lively stimulus in a lesson, Many other skills such as those listed in 'Towards a Minimum Entitlement: Brenda and Friends' (SSCR, 1984) are obviously used in these activities in addition to listening. Inclusion of these in the first year course at intervals, I believe, should enhance listening and enjoyment. However, it was important to try, in a limited way, to observe any improvement in my selected group of six. The following programme was adopted:

1. Observation of the group when engaged in the activity of listening to the teacher during a demonstration at the beginning and on two subsequent lessons later in the year.
2. Interviewing pupils in the selected group.
3. Using short comprehension written exercises.

My findings did show some improvement in their ability to follow instructions and in comprehension immediately after a listening activity. This was to be expected since the group was receiving an interesting or unusual stimulus. However, if these activities are to meet the aim stated at the beginning of this case study, then there should also be some significant improvement in other lessons where such a stimulus has not been given. After 1 year I cannot honestly claim that this is the case and would wish to have an independent observer to work with the selected group to determine the enhancement.

Some evaluation of the activities has been attempted with the help of colleagues and teachers in neighbouring schools. These stimuli are examples of strategies which are outside our normal teaching routine. A certain amount of preparation is inevitable but the overall cost of resourcing such activities is very low and the return in pupil motivation is high.

Fellow teachers of science may feel that such ideas are not science but more suited to active tutorial work. My personal view is that room should be made in our science curriculum in schools to allow skills such as listening, working in small groups, talking and non-verbal communication to be developed. Learning in science can only benefit from adequate preparation of pupils' study skills.

CASE STUDY 7: A DAY IN THE LIFE OF A GAS PARTICLE

Pauline Hoyle

Recently, a colleague Chris Laine and I were involved in curriculum development in which we were trying to ensure that science concept development was not hindered or affected by a pupil's lack of language development. We believed that, to be effective science teachers, we had to include language development strategies in our teaching and materials so that pupils had real access to science concepts.

We worked collaboratively to develop materials that enabled pupils to talk about their ideas before they were expected to read and write about them. To make this effective we developed collaborative learning techniques based on pupils working together in small mixed-experience groups of three to four.

We became quite adept at thinking up a whole range of activities that required pupils to talk about their ideas and teachers to listen to them. We also spent a lot of time and effort aiding and structuring any writing that we asked pupils to do. After a time we realized that we only asked pupils to do very restricted forms of writing and we questioned whether pupils really had only to write the 'traditional write-up' in science. We thought about why we asked pupils to write at all in science, though we knew that much of it was to do with keeping them quiet and making sure that the 'bright ones' had something to do at the end of the lesson.

We realized that pupils are already very good at writing stories when they come to secondary school and that we didn't use the skill in science. We also realized that writing can be used to help pupils think more deeply about a topic. We suspected that if pupils sometimes wrote in a form that was more familiar to them then they might possibly consider science more carefully.

We prepared materials for second year classes on 'Gases in the Air'. We carried out the standard tests for gases in the air, devised a cut-out to show the percentage of different gases, and then asked the pupils to write a piece on 'A day in the life of a gas particle'. They were given some help with getting started by viewing stories from a Sunday newspaper which were written about a day in the life of some person. They were also given an activity sheet (Fig. 7.1).

We found that most pupils attempted the task enthusiastically and the work was well above the usual standard. Some pupils were encouraged to

A day in the life of a gas particle

Write a story that starts: 'If I was a gas particle for a day what I would do . . . .
Use the following headings and ideas to help you write the story. Be sure to include all your knowledge
about how gas molecules actually move.

Where I went	How I got there	What I saw	What I did
— travel around the world in a cloud.	— got heated up and vibrated a lot.	— lots of enormous blood cells.	— moved about easily.
— visited Mrs Thatcher's house.	— got breathed in.	— lots of other gas particles around me called	— made friends with.
— made the Queen sneeze.	— bumped around with lots of other particles.	— other plant particles.	— went to the heart in some blood.
— inside a lung.	— got into a nose.	— Mrs Thatcher eating dinner.	— walked on the moon.
— rose up 60,000 ft.	— let the wind take me up.	— lots of uranium particles.	— visited the food house of the plant.
— visited the moon.	— got into the blood.	— nicotine on the lungs.	
— visited Greenham Common air base.	— got attached to a car or spacecraft.	— lots of germs.	

Figure 7.1

work together and produce just one piece of writing between them. This was particularly useful in groups that had bilingual pupils who need a lot of support in their writing. They contributed to the writing and thought a lot about the science yet didn't get turned-off by the difficulty of writing their ideas down on paper.

Some time later I was asked to record a video using these materials. It was thought that this would give other teachers an idea of how to include language strategies in science lessons, and how to include bilingual learners more fully in their science lessons.

It was decided to video a class of second year pupils with whom I had previously worked as a support teacher. The pupils were about to do the 'Gases in the Air' topic, so we videotaped a lesson of the pupils using a range of the material. The class teacher, Barbara Liscombe, had already set up collaborative learning groups and techniques with the pupils, so they were used to working in groups to complete tasks. They were also used to discussion work about science. I must emphasize that without this pattern already established, the pupils may well have found this activity difficult.

Each group was given different activities to complete. One group which included two bilingual learners, a dialect speaker and a mono-lingual were given the 'Day in the Life' task. The group was unimpressed the first time they were given this task to do, because they thought the other activities in the room seemed much more interesting. They didn't want to write a story. We got the usual excuse: 'We can't write Miss . . . can't we do that game.' Barbara and I decided to try giving the group a tape-recorder and let them tape their ideas. It wasn't that they couldn't write, it was just that they were reluctant writers who found the writing process quite difficult.

The tape-recorder initially gave them lots to argue about!

Barrie:	If I were a . . . , I would travel all around the world in a cloud. Visit Miss Margaret Thatcher's house.
Riccardo:	You can't *read* too much can you.
Barrie:	Just make an idea from this, Riccardo, just make an idea from this – go on. Here you are (passing worksheet to Kemal).
Lekon:	No wonder it's making that noise (the tape-recorder), switch it off and then it won't make any noise.
Riccardo:	Leave the thing alone.
Lekon:	Switch it off and then it doesn't make any noise. You see if it . . . if you switch it off it won't make any noise. Eh . . . stupid recorder.
Riccardo:	We can't cord/record on it 'cause . . .
Lekon:	No, you haven't switched it . . . , because I've got one

like this at home and when you switch it off it still works.
It doesn't matter.

After this initial problem the group settled down to negotiate what they were writing about and how they would do it. At this stage the idea of a 'rap song' started to emerge as each person took it in turns to write and say a line:

Riccardo:	Yeah, what are we doing? Tell me what you're doing? Tell me what you're doing? What have you done?
Darren·	Just, I would travel around the world in a cloud.
Riccardo:	That's yours, you read first. Kemal, man, organize the thing.
Darren:	Right, listen . . . (helping Kemal) . . .
Riccardo:	Who's reading? You read the . . . you read.
Darren:	I can't.
Riccardo:	Yes you can, yes you can, just read it.
Darren:	If I was a gas particle . . .
Riccardo:	Kemal, shut up!
Darren:	. . . for a day . . .
Riccardo:	You've got to speak up, you know.
Darren:	(Louder) . . . If I was a gas particle for a day what would I do? I would travel . . .
Riccardo:	I would . . .
Darren:	I would . . .
Kemal:	Travel around the world in a cloud.
Lekon:	You don't need that. All you have to do is . . .
Riccardo:	Visit Margaret Thatcher's house.
Darren:	Visit Margaret Thatcher's house and . . .
Riccardo:	Come on Lekon.
Darren:	. . . make friends, no, with . . .
Riccardo:	Lekon, are you doing this or what? Or are you doing your homework?
Lekon:	What?
Riccardo:	Are you writing?
Lekon:	Yeah, I'm gonna write in a minute.
Riccardo:	We're taping it, Lekon's not in it.
Lekon:	Are you taping it yourselves?
Riccardo:	We're taping it, you're writing it.

As the transcript suggests, peer pressure was brought to bear on Lekon to join in the group activity and stop doing his homework. Later in the video it was obvious that although Lekon was not very forthright in his ideas, he was, however, taking part in the activity, and not doing his usual opting out.

As the group discussion proceeded they considered what gases are in the air, how they would react in an explosion, what other uses the gases have. Unfortunately, there is no transcript of the discussion in the lesson when the group asked me 'How many oxygen would join with how many hydrogen in an explosion, Miss?' and such-like 'scientific' discussions.

The transcript shows that the group not only thought seriously about the science but they really supported each others' writing and reading:

Darren: If I were . . . oxygen I would live with fire. Come on we'll start it now.

Riccardo: Nitrogen, I'm opposite. I'll be nitrogen opposite of, I'll be nitrogen opposite of . . . (dictating to Kemal) . . . I'll, I'll be nitrogen the opposite of oxygen and put out the fire.

Kemal: I . . . (reading his writing) . . .

Riccardo: Just put that, it doesn't matter, you, you should know . . . I'll be hydrogen . . .

Darren: And put out fire.

Riccardo: . . . I'll be hydrogen and explode. I'll be hydrogen.

Kemal: How do you spell hydrogen?

Darren: H, I, just put H, I, G.

Riccardo: Oh, here it is (looks in book).

Darren: Just put H, I, G.

Riccardo: Hydrogen, H, Y . . .

Darren: H, Y . . .

Kemal: J, jer . . .

Darren: D, R . . .

Riccardo: No that's it D, it starts with D. No, G, that's it – Hydro . . . (all mumble) . . .

Riccardo: Explode . . .

Darren: I'll be hydrogen and explode . . .

Riccardo: . . . and explode . . .

Darren: . . . things up.

Riccardo: No, I'll be hydrogen and explode . . . (pauses thinking) . . . I'll be hydrogen and be very dangerous.

Kemal: Yeah.

On reflection, perhaps the most valuable thing we learnt from these lessons were that group writing does not in fact mean that individual pupils are not involved in the actual writing. On the contrary, all the group members were involved in the process of deciding what should or should not be put down. There was a sharing of the actual scribing but the confidence of all the group members was helped.

The use of the tape-recorder was invaluable. It enabled the group to go through the important writing process of drafting and re-drafting and yet maintained a 'good copy' of their final product. The tape-recorder doesn't require accurate punctuation or everything to be spelled correctly. It allowed the pupils to produce a product that they could be proud of. In following up this sort of activity, the pupils could also be encouraged to make a 'good copy' of their tape and so make them feel they were competent writers. In fact, a 'rap song' is much better heard than written, so in some ways these writers were in fact responding to audience and equipment very appropriately.

The group that were involved in this activity were predominantly bilingual. They were not fully competent bilinguals and they needed support in their writing. This technique gave them support and confidence to both explore the science and allow them to develop writing skills in English. As a result of using this activity on a number of occasions, we have now prepared further support materials, such as the beginnings of other pupils' stories and comic strips, which the pupils can sequence and then make a story around.

In many ways this activity was only a beginning. The success of the pupils' products inspired us to help pupils to write as a way of exploring their ideas in science. The important thing was to enable pupils to use different kinds of writing in science, to get them to think about what kind of writing format to use and when, their audience and, most of all, to *enjoy* writing in science.

CASE STUDY 8: WE WROTE A BOOK

Steve Whitworth

The beginning

Two years ago I shared the pleasures and pains of running a first year science course for low achievers. Although I'd like to believe that the course was an improvement on previous practice, it did highlight the considerable problems which have to be overcome when working with students with learning difficulties. Problems with retention of fact, the tendency to silly behaviour, difficulty with writing and, perhaps most seriously, the glimmering realization that they were different from the rest of the year.

This latter point was being unconsciously emphasized by the bureaucratic process of the school. At the end of the first year the school distributes most of these students throughout the rest of the year leaving a small group in the special-needs category. Because of an expressed interest and a degree of battle-hardening, I had the responsibility for teaching these 'remainders' in the second year of their schooling in science. I had no doubt that these students considered themselves as left-overs (do you remember the part of 'Kes' when they were picking football teams and Billy Casper was the 'remainder'?). As a first priority I had to lift the self-esteem of what was really a very nice group of seven students.

Working with Sue Dean (of the Write to Learn Project), we started to work on the relevant and fairly practical topic of food. We made a conscious decision to base our efforts on junior school practice; for me it was a break with my academic background and not an easy thing to do. However, once made, it was quickly seen to be the correct decision, because it laid more emphasis on the students' positive attributes and abilities. We were able to incorporate into the topic ideas which were designed to raise perceptions, to develop skills (rather than simply relay content) and to improve self-esteem.

We encouraged the students to become actively involved in the decision-making process. We often sat down and 'brainstormed' ideas. *We* listened to *them*. We made a 'bright ideas' notice board, where we displayed work of genuine value, for such it often turned out to be. We encouraged them to use flow-charts to express themselves, in order to get away from the rather turgid prose of standard scientific writing. We were quite happy to act as scribes, to let them see the genuine value of their thoughts without it being obscured by writing difficulties.

There were a lot of indications that this was a move in the right direction. The work progressed in a very satisfactory fashion, but I was a little unhappy that I was providing the direction, and I would have to provide more topics for them to do. In short, I was concerned about a lack of direction and a lack of foundation upon which we could build our combined futures.

The middle

At the same time as we were working with this group, Sue was also involved with a group of children in a local junior school. For some reason, they expressed a curiosity about what happened in 'science' in the 'big school'. Fortunately for all of us, Sue saw this as an opportunity to provide an impetus, a beneficial direction for the work we were doing.

The basic idea was quite simple; the special-needs group would act as teachers to the junior school pupils. They would try to enthuse, inform and direct them in their work on the Food topic. The mechanism by which this would be achieved was the production of a *book*. We saw this very much as an opportunity for the students to enhance their self-esteem by producing a piece of work of real quality and value. It had to be good, because they knew the students who were to use it were younger and therefore less able/knowledgeable than themselves.

We were making an effort to put them in the role of the expert; for them an unusual experience. They would know what they were talking about because they had also experienced the starting point. From this knowledge and experience they would be disseminators of knowledge, not receivers, again an unusual experience which we hoped would boost their confidence and self-esteem. Finally, in order to maximize the effectiveness of the learning phase, we encouraged them to work in pairs. If they worked alone for one reason or another, we made the proviso that they must communicate with other group members, because it was a group project. By this means we hoped that it would be possible to improve the social relationships within the group and encourage the individuals to see themselves as members of a group.

The *book* itself was nothing very special (at least to begin with); it was simply constructed of folded sugar-paper, slightly larger than A4, in a hard binding. It was encouraging that the pupils viewed the project with real enthusiasm and expressed a desire to fill it with *good* ideas.

We talked with the group about the basic structure of the book; they decided that it would be composed of topics, with worked examples to give the younger children something to aim at. They would include experiments and instructions on how to do them. They thought that it would be 'proper' science to do a survey of opinion and display the work in the form of graphs. This posed a rather special problem because they realized that this would produce difficulties for their readers. Accordingly, they not only produced a list of alternative examples for them to do but they also refrained from doing what they thought to be the 'best' example of survey work. One student wanted to do some drawings, another built up a collage of different food types. One made a record of the food which he'd consumed during the day. One pair had seen some work in the laboratory which was about food additives and wanted to do some research on labels.

Members of the 'topic-group' had to get the work to an advanced stage of drafting before we would consider final layouts or printing. Words had to be checked in the dictionary, alternatives sought and considered – we often heard snatches of discussions as to whether a certain word or phrase was suitable for their students.

In short, everyone was kept more or less continuously busy on one food-related topic or another for quite a long period of time. In fact, the work covered a period of 2 months, which sounds a lot, but at two periods a week it isn't really. The interest level kept surprisingly high, and I don't recollect any serious grumbling or murmurings in the ranks, perhaps because it was possible to treat the work in a modular fashion, each student pursuing their own interests.

The End – or perhaps a new beginning?

Because of the run-up to Christmas and the chaos caused by snow afterwards, we haven't at this time taken our book to the junior school (although a date has been set). In a sense, the product is less important than the process. The *real* value of the book lies in the insights it gave the students and their teachers about their abilities. It gave us all a confidence in ourselves and our group which has spilled over into new areas of endeavour.

The students (Some recollections from Sue)

Cara, Janet, Steven, Gary, Glynn, Jason (large) and Jason (small).

Cara: a diminutive blonde-haired girl who at the beginning of the year would avoid eye-contact and hide behind her bigger friend Janet. Just before Christmas was walking down the corridor with me explaining how she had interviewed several teachers for her 'What's your favourite drink survey?' and discussing how she should display her results.

Janet: always keen to please and do well, has been accepted by her male colleagues/peers in the group as an equal. She believes in herself and is prepared to make her contributions heard.

Gary: when first invited to write would resort to scribbles, now uses scientific language with confidence and has no qualms about putting his thoughts down on paper.

Glynn: always voluble is now more prepared to listen and to think through his ideas. He recently observed that it was good doing the book because everyone had worked as a team . . . (perhaps he included us in that team as well?!?).

Big Jason: usually coasts along, but really got terribly involved in the book. At times he showed remarkable understanding and maturity, especially in his spoken work.

Small Jason: enthusiasm is not a word one usually associates with Jason. His span of interest is rather short, but he did do some of the work and he didn't sulk too often.

Steven: often his enthusiasm ran away with his ability, but at least he was enthusiastic.

The book (Some recollections from the students)

We posed them this question at the end of the topic: 'What have you learned by making our book?' We got these replies:

1. It was important to work as a team.
2. We found it was quite hard to make a book.
3. We had to get our work organized.
4. We had to find out about food.
5. We had to make sense to other people.
6. We found the value of research.
7. It was fun doing something different.
8. We produced something good.

Acknowledgements

This article owes a great deal to my friend and colleague Sue Dean. I am grateful to her and for the work, under the aegis of the Schools Curriculum Development Committee, of the Wilts and Somerset Writing to Learn Project.

CASE STUDY 9: SCIENCE WITH RHYME AND REASON

Mike Watts

Pat D'Arcy, English Adviser for Wiltshire, has long tried to educate scientists in the ways of writing. Many years ago, at a conference of scientists in Birmingham, she convinced me of the importance of writing-for-learning for youngsters. She also introduced me to the possibilities of writing poems in science lessons. There is something about poetry (probably left over from my own schooldays) that brings on waves of self-consciousness, confusion and incoherence at first blush. Consequently, I'm usually to be found somewhere on the outer fringes of diehard philistinism. That said, in conquering some reservations, poetry in science classes has certainly been an interesting experiment each time I've tried it.

I have used poems in three ways. The first is the easiest: they are used to enliven worksheets, class materials and the otherwise dull paperwork of which lessons are often made. I took a leaf out of Keith Johnson's (1980) book *Physics for You*, which has now appeared in several editions. He litters each page with cartoons, jokes and jingles and has succeeded in producing a lively and popular text. My own bits of rhyme are of the awful groan-making pun variety but have the advantage that youngsters usually cannot take it for too long without retorting.

My first attempt was with a class of third years as we were looking at electricity and electromagnetism. Their follow-up work was entitled 'Ohmwork' and carried motifs of the 'joule thief who took a short-circuit on a mega-cycle' kind. This set in train a succession of bad jokes over the next few weeks, while one or two short verses began to appear in the homeworks being handed in. They were much more sophisticated than anything I had produced and the youngsters involved clearly enjoyed the time they spent in composition. Some were collaborative ventures and showed touches of craft in the way the words were put together. Poems (even limericks), however, were patently not everyone's cup of tea, and at this stage I never deliberately asked youngsters to write creatively in this way – I let it happen by default.

My second way of introducing poetry into science, however, was to change that. I chose two classes – a first year and a sixth form physics group. The first years were fairly straightforward and I suggested that if they wished they could present homeworks in the form of a poem. This produced poems on 'being a molecule', 'a day in the life of an atom', a 'ray of light', and so on. None were particularly outstanding but they did pave the way for discussions about science and scientists. Scientists are so often caricatured as humourless boffins and science as lacking in any aesthetic beauty or charm. Here we took time out to consider scientists as human beings and look at the interrelationships between science and art, music and literature.

In 'A' level physics we were looking at topics in electronics, taking a 'systems' approach to electronic devices, feedback and control. At one point I introduced them to the Japanese haiku. The haiku – as much as I know about it – is a fairly strict art form. The rules require it to be composed of 17 syllables, usually in the pattern of 5:7:5. It need not rhyme but must be evocative, encapsulating some image or emotion. We had earlier defined a system as 'an organised process for completing a task in a reliable and successful way' (after Pilliner and Snashall, 1987) and a haiku seemed to fit the bill as a system for writing short poems. This approach was fired by my reading of Hofstadter's (1980) book about Godel, Escher and Bach. Hofstadter teases the reader with haikus such as

Compressed poems with
Seventeen syllables
Can't have much meaning
and
Meaning lies as much
In the mind of the reader
As in the haiku.

The students picked up this element of fun and produced such ditties as:

Feedback systems
May shape and mould the signals
Which feed back, so that
and

Positive feedback
Makes continuous cycles
Which go on and on (and on and on and on. . . .)

I later introduced the group to the cinquaine, another idea I got from Pat D'Arcy. A cinquaine is a five-line poem with a form as follows:

first line: one word which must encapsulate the essence of the poem;
second line: two words which are verbs and express activity, the 'doing-ness' of what the poem is about;
third line: three words, all adjectives, capturing the description of what is going on;
fourth line: a four-word sentence which is the 'meat' of the poem;
fifth line: one word which sums up all the feeling in the poem.

The students fell to it with a gusto and produced such gems as

Physics
Twisting, turning,
Awkward, hard, exciting,
When will understanding come?
Tomorrow!

I have since used haikus and cinquaines with youngsters in a variety of circumstances and settings and usually with the same effect. I have started the lesson by introducing the formula and asking them to write something. Most groups have been very obliging and fall to task immediately – I always worry that if they were given a moment to consider they might rebel and refuse to embarrass themselves. Once the ice is broken, however, they enjoy what emerges and can quite easily be tempted to write another at the end.

My third way of using poetry is more circumspect. When tackling the topic of light with fourth years, I came across poems which illustrated some of the arguments in the history of scientific ideas (see e.g. Harvard Project Physics, 1968). Here I used the poems as forms of original writings to exemplify how the debate must have been conducted on the day. Newton's theories of white light as a mixture of colours, and the rainbow as a spectrum of colours, were not universally accepted by the intelligentsia of the time. His theories were in turns lauded:

Meantime, refracted from yon eastern cloud,
Bestriding earth, the grand ethereal bow
Shoots up immense; and every hue unfolds,
In fair proportion running from the red
To where the violet fades into the sky.
          (James Thompson, 'Spring', 1728)

and dismissed:

> May ye chop the light in pieces
> Till it hue on hue releases;
> May ye other pranks deliver,
> Polarise the tiny sliver
> Till the listener, overtaken,
> Feels his senses numbed and shaken –
> Nay, persuade us shall ye never
> Nor aside us shoulder ever.
> Steadfast was our dedication –
> We shall win the consummation.

This latter was by Goethe, who insisted that pure white was light's natural state and as such it could not be a mixture of other colours. This eminent German poet tried for many years to dissuade others from Newton's theory by a combination of unconvincing experiment and impassioned pleas. In class it was a way of delving quickly into the history of science, to set a historical context to ideas we take so much for granted these days, and to make the point that the great Newton did not always have everything his own way.

There has been nothing systematic in my use of poetry in science – I have tried it with different groups at different times and with differing degrees of success. My guess is that most science teachers would react to the suggestion that they try it too with incredulity, dire reluctance and not a little derision. Ah well.

> Then said a teacher, Speak to us of teaching.
> And he said:
> No man can reveal to you aught but that which already lies half asleep in the dawning of your knowledge.
> The teacher who walks in the shadow of the temple, among his followers, gives not of his wisdom, but rather of his faith and his lovingness.
> If he is indeed wise he does not bid you enter the house of his wisdom, but rather leads you to the threshold of your own mind.
> The astronomer may speak to you of his understanding of space, but he cannot give you his understanding.
> The musician may sing to you of the rhyme which is in all space, but he cannot give you the ear which arrests the rhythm, nor the voice that echoes it.
> And he who is versed in the science of numbers can tell of the regions of weight and measure, but cannot conduct you hither. For the vision of one man lends not its wings to another.
>
> (From *The Prophet* by Kahil Gibran)

CASE STUDY 10: ORAL ASSESSMENT IN SCIENCE

Brigid Bubel

This technique was developed by a group of teachers from four schools in the Portsmouth area, who were using *Science at Work* (1980) materials and wanted to develop a mode 3 course around it. Our target group, therefore, was the lower half of the ability range of 14–16 year olds, both boys and girls. For them, the main means of communication is, and will continue to be, the spoken word. As adults they are unlikely to want to write, but to talk to others about problems (and their possible solutions) at home/work or during leisure. Any ideas to do with science would be communicated orally. We hoped to encourage problem sharing, working out and testing solutions and so, as part of our continuous assessment course, we decided to include an end-of-course oral test. This would test three particular objectives:

1. To communicate scientific concepts.
2. To develop a comprehension of some basic concepts in science.
3. To develop an awareness of the contribution of science to the economic and social life of the community.

For each objective, we developed five criteria and marks ranging from 1 to 5 were allocated (see Table 10.1). Individuals then wrote questions based on different pieces of practical work mainly from the course. These were discussed by the group as a whole and modified if necessary, then tested with pupils of the same ability range as the target group.

We realized from discussion that, for two reasons, it might be difficult to decide whether a pupil could communicate a concept during an oral test. It could be that they had no idea of the concept(s) and, therefore, could not communicate it. However, it might also be that they knew the concept(s) but simply could not express themselves. We also expected some candidates to be able to weigh up the pros and cons of some issues and to come to an unbiased conclusion. However, many of our prospective candidates were not sufficiently well-read to be able to do this. To help resolve both problems, each candidate was given a passage to read before the oral which put forward some points for and against an issue to help generate some opinions. The passage was also read to the candidate at the beginning of the oral to ensure that it had been heard at least once!

Initially, pupils were videotaped during the orals and these tapes were then moderated by the group, using our criteria. We achieved a high degree of congruity, both in the initial group trials and later at workshops outside the group, after some questions had been discarded and others

Table 10.1 Assessment criteria

Objective 1: To communicate scientific concepts
5 Can explain his understanding of a concept in such a way that the listener can clearly comprehend what is being explained
4 Can explain his understanding of concept so the listener can comprehend but needs some prompting in later stages, *or* child communicates without prompting but listener only partially understands what is being communicated
3 Can explain his understanding to the listener who has some comprehension, but needs prompting
2 Has difficulty communicating at all without prompting
1 Has difficulty communicating even with prompting

Objective 2: To develop a comprehension of some basic concepts in science
5 Can recognize all the major factors of a basic concept in science and their relevant interaction
4 Recognizes some major factors and their interaction
3 Recognizes some major factors and has some idea of their interaction
2 Recognizes only some major factors without appreciating their interaction
1 Cannot recognize any of the major factors

Objective 3: To develop an awareness of the contribution of science to the economic and social life of the community
5 Ability to weigh up the benefits and drawbacks of science to the economic and social life of the community and attempts a conclusion based on the case for and against
4 Some awareness of benefits and drawbacks of issues
3 Aware of some of the economic and social effects of science on the community
2 Aware of some economic and social effects, but with particular bearing on the pupil himself/herself
1 Thinks of science in isolation from himself/herself and the community

---

rewritten and/or modified further. As a result of criticism during these latter trials we introduced open-ended questioning.

Organization

This section outlines guidelines for a group organizing science orals. It provides a sample timetable of events and some information for any group wishing to establish an oral exam.

Term before

1. *Standardization meeting*
This should be held by the group in the first school term. It should be led

by an experienced teacher who may be able (with the aid of our video!) to give some practice on the use of the criteria.

2. *Some notes on the use of the criteria*

(a) *Communicating concepts.* It is the ability to communicate which should be emphasized here, *not* the correctness of what is being said. Hence, the pupils need to be given some information and the pre-test passage has been developed to help them.

(b) *Comprehension of concepts.* This is in two parts: recognizing facts and understanding interactions. For example, recognizing a fact may be that the pollution of a river occurs because of a nearby factory; the interaction to be understood is that this is probably a major cause of fish death in that river.

(c) *Awareness of the contribution of science to economic and social life.* We consider this most important, hence our 10% allocation of marks. The pupil should be able to give a reasoned opinion of the benefits and drawbacks of a particular issue and *not* present bigotted ideas. Again, the pre-test passage is meant to help here.

Objective (b) is difficult to assess. The difference between 3 and 4 marks is not obvious at first but practice does help!

3. *Important arrangements*
The group will need to decide the following:

(a) Which experiments are to be used for the year's orals (at least two should be chosen).
(b) The date(s) (normally in the spring term) on which they will be held.
(c) Any external examiners needed – these, ideally, would be experienced members of the group and the number needed will depend on the number of schools taking part.

Remember to book the dates of the orals with your examination organizer at school, and that the spring term can be a very busy one for orals!

In the month before

1. A list of candidates should be drawn up with a provisional timetable and displayed in the appropriate place. (We allowed 10 minutes for each pupil.) Ensure you have a letter for each pupil to show their classroom teacher, during whose lesson they will be missing to have the oral.
2. Make sure any advance preparations are made for the experiments. Set up any longer-term experiments.
3. Ensure you have a timetable for yourself for use at the oral, copies of

Table 10.2 Science orals – timetable and marking grid

			Objective			
Date	Time	Pupil	1	2	3 Enter mark twice[a]	Total
e.g. 1.5.88	11.20	D. COLE	3	4	3 3	13

[a] Because objective 3 is weighted at 10, the mark obtained for this criteria should be entered *twice* as shown in the example.

the written passage, the Assessment Criteria, and a working tape-recorder for the day.

A week before

1. Give each pupil their letter with appropriate details.
2. Give each pupil a copy of the passages.

On the day

1. Have the experiments in the same room/laboratory and make sure they work/are showing expected results.
2. Make sure the tape-recorder is working and place it as unobtrusively as possible.
3. Remember you are allowed to ad-lib, so don't feel intimidated by our scripts. Take the opportunity when the external moderator comes to have a rest and listen to her delivery. Be prepared to compare and discuss your marks with the moderator.

After the orals

The tape-recordings will help to allay any fears you may have about your consistency of marking. You may re-mark if you find it necessary or if the moderator advises you to.

In conclusion

We have now used the oral examinations twice: 180 candidates were assessed on each occasion. With cross-moderation by visiting moderators from within the consortium, we reached good agreement on scores. The orals were quite straightforward to run and were generally well-received by the pupils. However, despite the introduction of the more open-ended questioning, we still felt that the orals were rather restrictive on conversation. For this reason, now that the course has been accepted as a mode 3 GCSE, we hope to develop continuous assessment of pupils, orally, through group work and, perhaps, by using work prepared by pupils in much the same way as English GCSE oral examinations.

Acknowledgements

I should like to acknowledge *Oral Assessment in Science* (Bubel *et al.*, 1987), completed by the Hampshire SSCR Group (08) of which I was a member, some of whose ideas have been heavily called upon here.

Appendix 1: Energy for the future – information page for teachers

1. *Passage to be read to pupils*

The sun is mainly made up of hydrogen gas. This gas is always being changed into another gas, called helium. During this change an enormous amount of energy is produced. Some of this energy reaches the Earth. This energy can be trapped by solar panels. Solar panels can be used to heat water in a house. They can even be used in countries where there is little sunshine.

The centre of the Earth is made up of very hot liquid rock. This warms up the rocks above. Water in streams under the ground can be heated by these hot rocks. Holes can be drilled into the Earth to find the hot water. This hot water could be used to heat houses.

The surface of the sea is always moving. A machine has been invented which can change this movement energy into electrical energy. At the moment, this is a very expensive way of producing electricity.

Windmills, too, are now being used to produce electricity. The energy in wind is free.

2. *Apparatus*

2 reservoir bottles fitted with thermometers
Plastic tubing
Wire loops
Pegboard
Black paper
Aluminium foil

Instructions: Set up the solar panels as on p. 31 of the pupils' booklet 'Energy' using lamps instead of sunlight. Make sure heat cannot be felt from lamp(s).

3. *Concepts Tested*

(a) There are many different forms of energy.
(b) The different forms of energy are interchangeable.
(c) Black surfaces absorb the light, shiny surfaces reflect light.
(d) Temperature change in the water is due to absorption of heat from the lamps.
(e) The longer the water is heated, the hotter it becomes.
(f) Fossil fuels are becoming very expensive so alternative energy sources need to be developed.
(g) Fossil fuels are running out.

4. *Benefits*

(a) Natural sources are being tapped, therefore cheaper.
(b) Environment will not be polluted to same extent.

5. *Drawbacks*

(a) Obtaining money to develop alternatives.
(b) Length of time it will take to develop them for use economically.
(c) How to overcome the problems of using natural sources of energy, e.g. when wind drops, when sun doesn't shine, etc.

Appendix 2: Questions asked by the examiner

Q.1   How much water is in this container? (Point to 'feeder' flask)
Q.2   Where is this water coming from? (Point to water leaving the panel)
Q.3   Look at the thermometers in each container. What is the temperature of each?
Q.4   What has brought about the temperature change? (concepts b and d)
Q.5   What type of energy does the lamp give out? (concept a)
Q.6   What type of energy does the water have at the end? (concept a)
Q.7   Why is the tubing covered with black paper? (concept c)
Q.8   Why is there tin-foil under the tubing? (concept c)
Q.9   If I (made this tubing longer)/(made more turns in this tubing), what difference would it make to the temperature of the water leaving the panel? Why? (concept e)
Q.10 Where would be the best place for a solar panel in a house? (concept d)
Q.11 Are there any drawbacks to using solar panels, say in a country like this? (Prompt: What about our weather?)
Q.12 Solar panels are thought to be one way of getting energy for the future. What is our main source of energy at the moment?

Q.13 What are the other alternative sources of energy for the future (apart from solar energy)
Q.14 Why do we need to look for/develop other energy sources at the moment (concepts f and g)
Q.15 What are the main benefits of these alternatives?
Q.16 Are there any drawbacks to them?
Q.17 Should we develop alternative energy sources? Why?

Appendix 3: Question analysis

Objectives and Questions

Questions 1 and 2 are not linked to the objectives.
Questions 3 to 9 test objective 2.
Questions 10 to 17 test objective 3.

Q.1 Is introductory to put candidate at ease.
Q.2 Is introductory to put candidate at ease.
Q.3 Requires only straightforward answers. It is important because it should ensure that (a) the experiment is working and (b) the pupil has actually seen that there is a difference in the two temperatures.
Q.4 May elicit a sufficiently detailed response to make questions 5 and 6 unnecessary, e.g. 'light energy from the lamp is converted to heat energy in the water thus raising its temperature'. This answer would probably ensure 3 marks for objective 2.
Q.5 May be necessary to ensure that the pupil understands the apparatus, if a good response has not been given to question 4. 'Light' is correct answer.
Q.6 May be necessary to ensure that the pupil realizes there is a difference in the energy given out by the lamp and that contained by the water, if a good response has not been given to question 4. 'Heat' is the correct answer.
Q.7 Is important to achieve higher marks in objective 2 and requires the idea of 'black surfaces absorb heat' better.
Q.8 Like question 7 for achieving higher marks in objective 2 and requires the idea of 'light coloured surfaces reflecting heat better'.
Q.9 Like questions 7 and 8 is important in achieving high marks for objective 2. It requires the pupil to understand that the temperature of the water leaving the panel would be greater because it would have been heated for longer inside the panel.
Q.10 Requires answer 'somewhere where as much light as possible would shine on it'. Such an answer would ensure that the pupil had grasped concept d. The 'why' part of the question may be

unnecessary. If the pupil has no idea of the 'correct' answer, it is probably indicative of a score of 1 for objective 3.

Q.11 May require this prompt if the weather has been unusually consistent just to remind them that it is normally rather changeable. The question should elicit answers such as 'The sun doesn't always shine brightly so unreliable', etc.

Q.12 Should elicit 'fossil fuels' or name 'oil'.

Q.13 Should elicit some covered in the course, e.g. wind, waves, geothermic.

Q.14 Should elicit idea of fossils becoming very expensive as they become scarcer. This will ensure concepts f and g have been grasped. It would also probably ensure 3 marks for objective 3.

Q.15 Refer to lists of benefits and drawbacks. One of *each* of these or any.

Q.16 Suitable others for 4 marks for objective 3.

Q.17 Good answers to this will ensure 5 marks for objective 3.

SUMMARY

To summarize, we consider our usual four questions:

1. Which aspects of active learning do they encourage?
2. What planning do they require?
3. What organization is needed?
4. What else is important?

1. Which aspects of active learning do they encourage?

The value of talking and writing in science lessons, says Carré (1981), is that children learn to interpret their observations. He maintains that by talking through scientific ideas, and forming their own meanings, children are better able to learn science. Discussion, as case study 7 indicates, enables youngsters to organize themselves, make decisions and display their understanding. It also provides good opportunities for youngsters to evaluate themselves and others, as case study 6 shows. Many parts of active learning, from our checklist, feature in discussions and small group work. Taking the focus away from the teacher and making individual contribution an important aspect, shifts the balance towards youngsters taking responsibility for their own learning.

It certainly provides excellent opportunities for direct skill teaching and space for youngsters to direct and own their own activities. If structured carefully, it can be much less threatening than the whole-class group where some can feel out of their depth, or be remote and unchallenged by what is going on around them.

2. What planning do they require?

Aspects of planning are dealt with in our next two sections. Here we think about just one case study. Steve Whitworth's example (case study 8) obviously requires a great deal of forward planning – in making links with primary schools, encouraging pupils, working at *their* pace and being ready to anticipate many different demands for investigations and re-source materials. Effective writing tasks such as in case study 8 may well require prior discussions with English department colleagues. Case study 9 also would indicate that inter-departmental cooperation is very valuable as a forward planning strategy.

3. Organization

We return here to our earlier framework of 'prior conditions' and draw out some aspects for each of the case studies. The first stage lies with the teacher's own implicit ideas about discussion. Just as Mike Watts first required pupils to examine the process of discussion, so teachers need to consider for themselves a set of criteria for what constitutes a good discussion. They may well, for example, be centred around cognitive change, personal development or social interaction. Whatever the criteria, they need at some point to be made explicit if the aims of the lesson are to be achieved and evaluated. The criteria may well be different in different lessons: case study 5, for instance, deals with criteria for listening skills whereas another lesson might feature group cooperation.

Case study 5 provides an example of methods that teachers might use to ensure each of McClelland's (1983) conditions for discussion are present. Mike Watts began by using 'problem situation' cards depicting gravity. He could equally well, though, have had the youngsters generate their own problems, or have used problems about the processes and procedures youngsters had met in a practical situation.

McClelland's second condition was that of status – the difference in authority between teachers and pupils. However, it might mean some-thing different in this context. Discussion, as a learning activity, is often viewed by youngsters as 'not doing any real work'. That is, work is something that produces only a written output. So a lesson spent in discussion can mean there is little to show for it and the task of discussion can often be construed as time-wasting. Certainly this is a problem that might have arisen in Mike's first lesson, given that half of it was given over to discussing discussion! However, Mike seems to have slipped by this, by ensuring that youngsters understood the aims of the task very clearly and had a particular objective to fulfil by the end of the time. In this case they were asked to make a report back. In case study 6 the youngsters might have been asked to assess each others' listening skills. Another useful task is to have youngsters focus on the processes of what happened in the discussion and complete a piece of homework on how they personally might influence the process next time so that the discussion moves in a different direction.

McClelland's meaning of status, though, is in the difference of authority between teachers and youngsters. Numerous studies show how pupils 'collude' in getting answers from teachers to 'get through the lesson'; how teachers are called in to 'arbitrate' over a point of contention when the youngsters are in small groups; how teachers – often unwittingly – use leading questions to get right answers, or simply use their authority to assert their own, final opinions.

This all reinforces the teacher's role as expert. There are circumstances where one would want this to be the case. One would also want youngsters to feel they could challenge the teacher's ideas in much the way they would challenge the ideas of others. In case study 5 teachers added their ideas to the list after the pupils had reported back, thereby giving equal status to theirs and those from the class. Another way might have been to ensure that discussion activities are governed by a negotiated framework of rules. Group members are then free to challenge each other, and the teacher, for infringements of the rules. This tends to change the status of the teacher to that of a referee. Some useful starting rules are:

- criticize the argument, not the person;
- ask for explanation if you do not understand what is being said;
- try to base the arguments firmly on the problem in hand;
- listen carefully to others and give them time and space to lay their reasoning before you. Interruptions before they have developed their ideas are not acceptable.

Group size was another of McClelland's (1983) prior conditions. In collaborative group work, group size is obviously crucial. They must be of sufficient size to allow for ideas to be productive, but at the same time provide everyone with a role to play. Four is often sited as an optimun number but, as the case studies indicate, pairs, threes and sixes are used successfully. The number of groups can be reduced by techniques such as 'pairing and sharing', or by dividing the class into groups of four to begin a discussion session with a 'brainstorm', and then selecting two or three of the most profitable ideas to pursue. Another tack is for the teacher to conduct the discussion with one small group while the rest of the class act as observers. Each observer is given a skills list, asked to observe a specific person and provide her/him with feedback later. This could be used to follow-up Harry Moore's lesson on listening skills in case study 6. Similarly, observers can analyse the process of discussion, i.e. which ideas moved discussion on, what sort of questioning occurred, and so on. It is a mechanism Mike Watts might have used in a later lesson, after the youngsters had developed criteria for a good discussion.

Large class discussion often proves to be very difficult unless con-ducted as a formal debate. Returning to case study 5, it requires a high degree of questioning skill, and a keen eye for managing the contribu-tions, to enable all youngsters to have the opportunity to participate if they want to. As case studies 6 and 7 show, though, it is often more useful to reduce the size of the groups. It has the advantages of allowing more youngsters to participate in the time available and of providing a less threatening experience. Indeed, this can be an important rationale for many types of small group work. Throughout all the chapters of the book small group techniques are used to add variety to teaching approaches. They allow for a high degree of active and collaborative learning. They are particularly useful when faced with classes containing a wide variety of fluency in English. Creating small groups in which there is a carefully chosen spectrum of fluency and understanding and encouraging dis-cussion in mother tongue, with report-back in English, allows youngsters to both understand and learn.

4. What else is important?

Understanding the processes of a discussion might help youngsters to participate better. In this last part of the chapter, we examine each of the steps in that process and suggest ways in which teachers might facilitate each one. We have to assume that the prior conditions outlined in section 3 of this summary have already been met.

Getting started – small group situations

Dividing the class into groups is an obvious yet very important first step. Teachers need to make it clear that they are the overall organizer, and that group composition is decided on the basis of ensuring everyone gets a chance to work with everyone else at some point during the year. This implies that groups will have different compositions throughout the year, taking account of such things as levels of verbal participation, listening skills, gender, race, friends, enemies, depth of knowledge, and so on.

The next step is to ensure groups know exactly what their task is, what objectives they have to complete and how long they have to do it. This is important whether the technique being used is discussion or writing. Sometimes, as in case studies 5 and 7, this is done verbally, sometimes in a written form, as in case study 6. The group might also need reminding of the rules that have been established in the class with regard to discussion – allowing others time to reflect and speak, etc.

Structuring discussion

Sometimes, as in case study 6, it is important to focus youngsters' attention on discussion itself. There are several ways, for example, to give 'procedure cards' which provide a set of instructions such as:

- one person states the topic of discussion;
- person A gives an opinion;
- person B gives an opinion;
- other youngsters ask for reasons for putting forward those opinions;
- A and B are invited to give evidence in support;
- the whole group sorts out any discrepancies.

Occasionally, detail someone to observe what is happening to the discussion in a group without taking part. They can then report to the others on aspects which often escape notice, for example whether it is always the same person's ideas which are adopted, or whether girls or boys dominate. Once the work has begun, the teacher's role is ambivalent. Are they just there to sort out disputes or is there a more positive role to play? It certainly helps to move around groups, to stop the discussion and ask for a summary. This helps pupils to practise summarizing, to see who has been listening, and allows the teacher to ask further questions to move the discussion on.

Ending discussion

The easiest way of ending is by means of a report-back. It helps to structure the reporting session carefully. For example, if the reports are verbal ones, constrain the time for reporting to make the session manageable. By contrast, written report-backs can be very valuable. In case study 8, the purpose of the whole activity was the 'report' – the book being written. They can involve individually written reports, e.g. for homework, in the form of descriptions, poetry, imaginings or many other mechanisms. Alternatively, collaborative reports, such as a poster session, are useful because they allow groups to refer back to earlier ideas, but they do require careful organization. The planning for the report-back session should always ensure that groups have time to write their own posters and read those produced by other groups. This poster session ideally needs following up with a plenary session, involving the whole class, which results in a discussion of all the ideas produced, and comparisons between them. Reports should also focus on the discussion itself as well as the topic of the discussion. Again, it is profitable to ask one youngster to act as an observer for each group, or as a 'roving reporter' for the whole session, in order to examine the processes of discussion. This concentrates on what they saw, not what people talked about. They may focus on the roles played by individuals within the group – who summarizes, who dominates, who opted out, who facilitated, etc. It is not a report-back that is required often, but it does help to raise youngsters' awareness to the fact that the composition of a group can change the roles performed by people, and that the roles adopted are within their own control. They do not always have to dominate or play devil's advocate, for example.

We leave the last words to Clive Carré (1981):

Teachers who see the potential of language for assisting learning and realise that writing and talking can act as part of a connection-making process will therefore:

—encourage pupils to put new ideas into words and talk about them to the teacher and each other;
—provide ample opportunities for pupils to apply their knowledge and to re-sort new information in old and new contexts;
—suggest different ways of writing to clarify understanding.

BIBLIOGRAPHY

Assessment of Performance Unit (1982). *Science at age 15. A Report.* Department of Education and Science. London: HMSO.

Association for Science Education (1981). *Study Series Six. Language in Science.* Hatfield: ASE.

Bubel, B., Benning, M., Cole, D. and Murphy, P. (1987). *Oral Assessment in Science.* Hampshire: Hampshire LEA.

Button, L. (1974). *Developmental Group Work with Adolescents.* London: University of London Press.

Carré, C. (1981). *Language Teaching and Learning: Science.* London: Ward Lock.

Devon Group (1986). *A Skills Approach to the Teaching of Science in the Lower School.* Exeter: Devon County Council.

Gibran, K. (1980). *The Prophet.* London: Heinemann.

Gilbert, J. K. and Watts, D. M. (1983). Concepts, misconceptions and alternative conceptions: changing perspectives in science education. *Studies in Science Education* 10, 61–98.

Harvard Project Physics (1968). *Text 4, Light and Electromagnetism.* New York: Holt, Rinehart and Winston.

Hofstadter, D. R. (1980). *Godel, Escher, Bach: An Eternal Golden Braid.* Harmondsworth: Penguin.

Johnson, K. (1980). *Physics for You*, O-level edition. London: Hutchinson.

McClelland, G. (1983). Discussion in science lessons. *School Science Review* 65 (230), 129–33.

Pilliner, G. and Snashall, D. (1987). *Systems Electronics.* London: Macmillans.

Science at Work Series (1980). Taylor, J. (director). London: Addison Wesley.

Secondary Science Curriculum Review (1984). *Towards the Specification of Minimum Entitlement: Brenda and Friends.* London: SSCR.

Watts, D. M. (1982). Gravity – don't take it for granted. *Physics Education* 17, 5.

Watts, D. M. and Gilbert, J. K. (1984). *Appraising the Understanding of Science Concepts: Gravity.* Mimeo. Department of Educational Studies, University of Surrey, Guildford.

4: PROBLEM SOLVING

INTRODUCTION

Problem solving is fast becoming one of those expressions that mean all things to all people. It is sometimes difficult to decide if it is a serious, responsible approach to teaching and learning, or if it is simply another educational bandwagon. It has been discussed at length by the Assessment of Performance Unit; it tops the list of recommended teaching approaches in the Department of Education and Science's policy statement; it is firmly in the focus of technologically and vocationally orientated courses, from the Certificate of Pre-vocational Education (CPVE) and CDT to Home Economics; it is advocated for primary science by the Engineering Council, and is now to be found as the subject of in-service seminars and conferences. Problem solving, as an educational activity, is not new. There is a large and detailed body of research that extends over many decades, exploring the characteristics of school problem solving and problem solvers: we list some of it in the references for this chapter. What *is* new is that the type of problem under consideration has changed, and tackling these new problems is becoming acceptable, respectable, classroom practice. It is worth discussing these two points separately, allowing us then to put the four contributions in the chapter into a clear framework.

There is enormous variety in the types of problems seen fit for solution and there have been several attempts to categorize problems. They have sometimes been classified as 'open' or 'closed', 'formal' or 'informal', as more or less 'curriculum dedicated' (Munson, 1988). In practice, these range from 'IQ tests' to 'egg-race' activities, from puzzles to 'real-life problems'. For instance, consider the following list of problems directed at youngsters in the 11–18 age range:

1. Make a 'helicopter' from a piece of card 14×14 cm to spin as far as possible when flicked by a finger from a table top.
2. You are lost in the wilds of a tropical country. You have not drunk anything for 3 days. All you have around you is some swamp water, some coconut trees and some bamboo trees. You have a sharp knife, some matches and a spare shirt. Find a way of producing pure water from the swamp. You must also find a way of proving it is pure.
3. Given $f = ma$, where a is 4 m/s/s and m is 3 kg, find f.
4. Why doesn't this typewriter, tape-recorder, etc. work?
5. What can be done to help infirm or elderly people who find difficulty in pouring boiling water, operating taps, cooker controls, etc.?
6. How can someone be helped to reduce their heating bills?
7. What is pollution? How does it arise? What are the effects of pollution? How is it detected? How is it tackled?
8. What are the benefits, effects, disadvantages of colour and flavour additives in food?
9. You are part of a family of five planning to buy a new car. Using technical specifications, draw up the criteria for choice and then choose an appropriate car.

Some of these can be seen to be relevant to a traditional science curriculum, some are described as open-ended, some – particularly the latter ones – are listed as 'real-life' problems. They are all, however, seen as problems to be solved.

This list does beg the question of how we might define a 'problem'. For us, a person has a 'problem' when she or he has a goal which cannot be achieved directly. Kahney (1986) makes a distinction between well-defined and ill-defined problems. Well-defined problems are ones in which the goal, the possible moves (possible routes to a solution) and strategies are all given at the start. Ill-defined problems are ones in which the goal and the permissible moves have to be supplied by the problem solver. This distinction is useful as far as it goes: it separates those problems and instructions which are given to the solver by someone else, from ones which the solver thinks up for him- or herself. For us, the terminology seems unduly pessimistic, and we would want to add a third category. Our distinction would be between *given* problems, where the solver is given the goal and strategies, *goal* problems, where the solver is given the goal and nothing else (they have to decide and develop their own strategies), and *own* problems, where solvers decide both the goal and the strategies. These are not meant to be hard and fast categories but to be consistent with the underlying theme of the book. For us, the central virtue of problem solving is that it is a means of transferring some of the responsibility for learning to the learner. The main point of adopting the approach in schools is that the emphasis is on the learner using a planned approach (their own or someone else's) to tackle a problem. It becomes their responsibility to delineate the problem, decide on what an appropriate solution might be, derive and test possible solutions, and choose the point at which they think the problem has been solved.

In this sense, problem 3 in the list above is very much a *given* problem. The goal is specified (find '*f*') and, while all the permissible moves are not explicit in the problem statement, they are clearly defined elsewhere (probably in the preceding pages of the textbook, or in classroom notes). It is not permissible, for example, for the problem solver to provide the moves and, say, to juxtapose the two figures and write 43 or 34 as a solution. It is also very much an oddity in the list. The bulk of the research literature we mentioned earlier refers to this kind of problem. For example, the Newell and Simon (1972) theory of problem solving focuses primarily on puzzles, IQ type questions and text book calculation problems. Past emphasis in science education has been on, for example, solving chemical equations (Ashmore *et al.*, 1979) or doing molarity calculations (Black and Elliott, 1982).

The shift in school problem solving has been away from this *given* type of problem towards what we have labelled goal and own types, i.e. to the other types of problems on the list. The problem might be one that is presented to the solver, or one generated by their own thinking, actions or lifestyle. It is a welcome change. *Given* problems often seem spurious and, while not denying that they sometimes have an important role to play, they bear little relevance to students' normal everyday life. For the problem to be meaningful, the goal has to be worth achieving, the student has to own the problem.

The adoption of problem-solving activities in school science courses is happening quite slowly. Early work in schools (e.g. Mathews *et al.*, 1980) was a useful attempt to simulate and explore the roles of scientists 'doing' science. It is increasingly seen as a valuable way of providing more open learning situations where youngsters are less constrained by didactic teaching methods. One particular growth has arisen in the twilight zone between science and technology. Paul Black and Geoffrey Harrison (1985), for example, see it as the main element in the area formed by the complementary overlap of the two subjects. They see science and technology being resource areas, both conceptually and physically, to be drawn upon in the service of solving problems. From another direction, the officers of the Further Education Unit (1986) see problem solving as a principal means of meeting many of the aims and objectives behind the Certificate of Pre-vocational Education (CPVE). They too wish to draw on the resources of science and technology:

> One way of considering science and technology within the CPVE context is as 'real world' problem solving. To this end, students

need to attain basic skills in the application of scientific method and technological processes to solving problems.

The kinds of problems they have in mind are clearly more open-ended and relevant to real-life situations than the others. These are the kinds of problems such as numbers 4–9 in the list above. There is no reason, though, why these kinds of problems should be reserved for the 'vocational' training of the 14–19 age group. They seem perfectly good problems, as Beverley Cussans illustrates in her contribution, to set youngsters at many different stages of their education.

All this has several implications for the management of problem solving in schools. Do you set the same problem for a whole class group so that they generate different solutions? Do you set the same problem to all but ask different groups to tackle different aspects so that you build up a composite picture at the end? Do you set different problems for different groups, or even a different problem for each individual and allow them to work separately? Before beginning problem solving, do you first deliberately teach all the facts, concepts and skills so that the pupils will have all the relevant information at their fingertips? Or do you use the motivating power of problem solving as a means of getting the pupils to decide and satisfy their own knowledge needs?

Some of the problems have 'hardware' solutions (the production of a tool or artefact), whereas some problems, like re-routing traffic to avoid noise near a hospital, or choosing a new car, have 'software' solutions. This means the pupils will need to have recourse to the use of laboratories, workshops, libraries, computers, the great outdoors, etc. In this sense, teachers really must begin to develop practical alternatives to teaching science and technology. The CPVE board, for example, suggest (FEU, 1986):

> Safety considerations mean that an appropriately trained teacher needs to be in attendance in laboratories and workshops, but it does not necessarily mean that all students should be doing the same thing at the same time. The concept of 'open access' or 'drop-in centres' or 'resource bases' has been used, enabling students to visit the supervised laboratory or workshop as they wish.

Our contributors in this chapter embrace many of these issues and, hopefully, do not so much present answers to the questions above as provide some ways in which they have tackled problem solving successfully. All four have, at some point, been part of the Secondary Science Curriculum Review, and we are very grateful to SSCR and the work of the groups in Cambridgeshire and Kent. Andy Howlett looks at a 'Desert Island' problem with first and second years, and provides some general considerations about the value of problem solving and the changing role of the teacher. Beverley Cussans' groups tackle 'Pollution' and her case study raises interesting questions about the management of problem solving in school laboratories. The problem described by Dave Wallwork is very much a 'hardware' one, where youngsters worry about how a blind person might safely pour boiling water. He also describes a novel approach to timetabling, where normal lessons are suspended for the day. Mick Nott takes a different but interesting slant, introducing a strong multicultural context within which the pupils are asked to work, and focusing particularly on the reactions and feelings experienced by the teacher during problem-solving activities.

Douglas Adams' (1982) hero in *Life, The Universe and Everything* argues that some problems can conveniently be labelled an SEP (Someone Else's Problem), whereupon it can be totally ignored, becomes invisible, and will eventually go away. In this way, by making it an SEP, he gets rid of an unwanted spacecraft from the centre of the pitch in a test match at Lord's. In a less humorous vein, if youngsters are not engaged in setting the goal and working out the routes and strategies for themselves, they will see the goal as not worth achieving. Then the problem disappears – it becomes an irrelevance. Problem solving in school science is worth a book (several?) all by itself. All we can do here is to offer a sample and flavour of what it entails. Hopefully the bibliography will allow the interested reader some entry into the wide variety of work that has been achieved around the country.

CASE STUDY 11: THE DESERT ISLAND PROBLEM

Andy Howlett

'Problem solving' has been used in science education for many years, mainly as a vehicle to enable teachers to evaluate students' ability to transfer concepts and understanding from the classroom to real-life situations. With the onset of a more process-based science curriculum, however, 'problem solving' has taken on a whole new meaning. It has become the embodiment of the 'scientific process', the 'real method' of investigation.

The development of topics and investigations for problem solving has been taken up by several groups of teachers and researchers. One such group, at Peterborough, a part of the SSCR nationwide series of science teachers' working groups, have developed, trialled and evaluated a series of investigations based around the theme of survival on a desert island (Peterborough SSCR Group, 1986).

The desert island problems are designed as a resource-based investigation in science, involving 11–13 year olds in the study, design and solution of real-life problems. (For examples see Appendices 1 and 2). The problems, and the resources that go with them, have been designed for students of all abilities (there is a system of individual clues, etc.) as well as interlinking the social and ethnic background of all students. The work is based on small groups and, therefore, can be used with streamed, banded and mixed-ability groups, although it is, in essence, for mixed-ability groups.

By definition, the students undertaking the desert island problems are, within limits, the organizers of their own learning. Through interaction with a series of resources and investigations, the students research, experiment, evaluate and realize solutions to the problems. In doing so, they encounter and acquire the skills and concepts encompassed by that area of the curriculum. This form of learning strategy is cross-curricular, encouraging students to research and use expertise from all areas of the school – from library skills to CDT and home economics. The problems have evolved from the desire to produce a problem-solving approach to the learning of 'Energy Topics' within the lower school.

The resources available on the imaginary desert island are minimal, but students are encouraged to use standard laboratory apparatus to help test their designs to predetermined specifications. The main sub-sections of the problem are:

1. Production of drinking water (see Appendix 2).
2. Shelter design.
3. Waste disposal.
4. Heat.
5. Cooking facilities.
6. Food production.
7. Clothing and insulation.

In the classroom we have organized the youngsters using one of three distinct implementation strategies. These are mixed-problem groupings, single-problem groupings, and single-problem classes. In the first two cases all seven sub-problems are solved within a single class of, say, 28 students. The first approach requires students to be grouped in units of seven so that each student can work on an individual area of the problem and report back to the group. It requires the groups to work as a team in order to solve the entire survival problem. Group reporting sessions are encouraged. Individuals develop skills through being required to study all areas of the problem.

The second approach requires the class to work as a unit in the same way as the groups in the first approach. Students are divided into seven groups and each group is charged with the task of solving one of the seven sub-problems. Each group then reports back to the class as a whole on their progress through their part of the work.

The third approach requires the whole class to tackle each problem in turn. The equivalent to the 'class practical', students work in pairs and then compare each other's outcomes and findings. Each approach has been found to have individual benefits, usually depending on the personal preference of the teacher.

All the three suggested strategies require students to work in teams following their own plans through a set series of processes. These processes are constrained by the use of 'Apparatus order forms/Assessment sheets'.

An 'open ended' style of problem solving requires a change in the teacher's role from traditional science lessons. The teacher has two functions in the laboratory during this project:

1. *As controller and assessor*: this represents a role very similar to that adopted by science teachers for some years, that of professional advisor to assessment, evaluation, discipline and control.
2. *As a resource*: this is a dissimilar role to previous traditions where the teacher acts as a resource to the student-centred learning that is taking place. The teacher is neither the major nor the exclusive resource but is classed as one of many important sources of help that the students can try when in difficulties.

The differences between this role and that for a didactic approach is shown in Table 11.1. At the end of the topic the student and teacher evaluate the work through the use of questionnaires.

The students use the teacher as a centre for appropriate clues and hints (through the use of 'clue cards') and as a regulator and assessor for their plans and designs. The major part of the teacher's work is in preparation.

Table 11.1   Teacher and student roles for the didactic and problem-solving approaches

Didactic	Problem solving
Role of teacher[a]	
Initiator of work	Responder to ideas
Organizer of experimental work	Adviser to experimental design
Chief supplier of information	Resource of information (one of many)
Evaluator of results/work	Adviser to evaluation
Role of student	
Receptor of information	Researcher of information
Guided experimenter	Experimental designer
Minor evaluator	Major evaluator
Responder to ideas	Initiator of work

[a]Other roles (assessor, safety, control, etc.) are common.

This type of approach requires thorough consideration of content, re-sources, equipment and possible clues as well as being forewarned of possible stumbling blocks for individual students, and blind alleys to investigations through lack of equipment and inappropriate resources.

Evaluating the approach

We have used this kind of system extensively in lower-school science in several Cambridgeshire schools, producing excellent results in terms of both learning and understanding. Our trials have been evaluated using data collected through checklists and questionnaires.

Data collected from the checklists show that students experience a wide range of skill and process learning, ranging from the correct use of Bunsen burners to ray optics and heat transfer. Although the factual content can vary due to the nature of the course, the areas of experience of the student can be constrained by the clues and apparatus given.

Our analysis of both student and teacher questionnaires showed that students were definitely motivated by problem solving. Students who encountered this type of work for the first time felt that group work was something that needed practice. Interestingly, in practice, inter-group help was rare. The large majority of students found the resource-based approach to the investigations an exciting and rewarding method. Teachers were convinced of the high level of learning and understanding that was taking place.

In all cases, there was a dramatic increase in the use of school libraries by students in order to research ideas and information. This trend did not cease with the project but has continued after the event. Students found it difficult initially to adjust to the role of the teacher-as-resource rather than leader, but very soon came to rely on themselves and their own resources rather than the teacher.

The work certainly increased student motivation and participation. With any large group project there is always some possibility of 'hangers on' within a group, but in actuality this has been a very rare occurrence. The students have also encountered and learned far more than would have normally been covered in the same period of time but the coverage is not uniform across the whole class and so some 'blocking in' of content areas, discoveries and understanding is necessary. The approach has increased the use of cross-curricular courses in all the schools which have been using it, as well as producing an increase in the active use of the library by students during and after the project. It has enabled students to bring their full prowess to bear as 'investigative scientists' as well as linking past experience, the real world and the laboratory together.

No innovation is without its drawbacks, but in this case these are ones that can be overcome. To the inexperienced problem solver the work can be quite tiring (for both staff and students) due to the intensity of concentration required. This has been shown to become less of a burden with practice and most teachers and students become accustomed to the hard work.

As an active use of the 'scientific method' it is a viable and effective tool. It is also of tremendous use in putting science into its wider social and political content. But above all, it gives the centre stage to the student, continually encouraging him/her to become an active investigator and evaluator – a major aim for science education fulfilled in one strategy.

Appendix 1: Stranded

You were in a plane that crashed. Fortunately, you have survived and have reached this uninhabited island. In the lifeboat you have a small amount of food, an axe and a First Aid Kit.

Where on the island is the best place for you to stay for

(a) a short time (less than 1 week)?
(b) a long time (more than a year)?

There are four possible sites. Which of these is the best? Can you find a better site? There are four grades:

Very convenient	4
Convenient	3
Inconvenient	2
Impractical	1

Work out the score for each situation according to its convenience to your needs on the island.

1. Water		River/lake
2. Food	Goats	Plain/hills
	Fish	Sea/river/lake
	Coconuts	Coast
	Fruit	Forest/hills
	Crops	Plain
	Wild Fowl	Plain/lake
3. Habitation	Hut — wood	Forest
	grass	Plain
	mud	River
	Boat	Shore
4. Clothes	Goat skin	Plain
	Deer skin	Forest

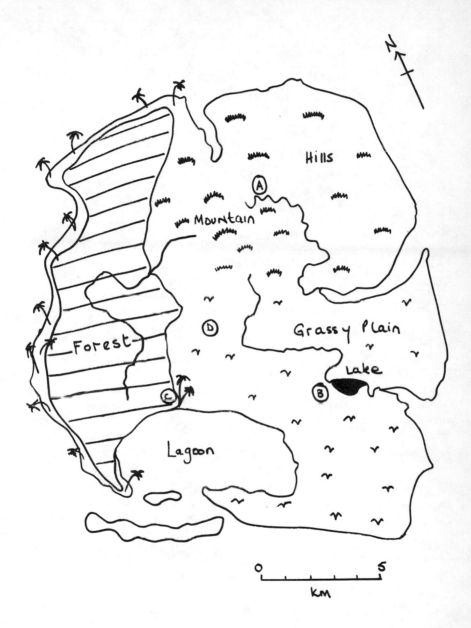

Some articles may be unnecessary, e.g. clothes on a short stay, score these things O.

		A	B	C	D
Water					
Food	Goat				
	Fish				
	Coconuts				
	Fruit				
	Crops				
	Wild Fowl				
Habitation	Hut				
	Boat				
Clothes	Goat skin				
	Deer skin				
Total for each site					

N.B. Discuss and give values to other factors, e.g. mosquitoes in the marsh, etc.

Appendix 2: One desert island problem – drinking water

Problem: You are lost in the wilds of a tropical island. You have not drunk anything for 3 days. All you have around you is some swamp water, some coconut trees and some bamboo trees.

You have a sharp knife, some matches and a spare shirt. Find a way of producing pure water from the swamp. You must also find a way of proving it is pure.

Equipment used:	Filter funnels, filter paper, conical flasks, distillation tubes, condensers (for extension), beakers, pond water
Examples of clues:	Picture of steaming kettle; picture of filter coffee maker; diagram of thermometer
Other resources:	Anglian Water Authority posters on water treatment (free)
Extension problems:	Separating oil and water; finding out what is in pond water
Approx. time required:	4–5 hours
Areas covered:	Filtration; distillation; temperature measurement

CASE STUDY 12: PROBLEM SOLVING – A WAY OF
PROMOTING SCIENTIFIC SKILLS AND PROCESSES

Beverley A. Cussans

In recent years the science staff at the Geoffrey Chaucer School, Canterbury, have been working together to produce a new science course for pupils in years 1, 2 and 3. The aim is to produce a balanced science course for girls and boys of all abilities, with the main emphasis on developing scientific skills and processes. The course is called *Foundation Science* (Cussans and Munson, 1987).

The writing of the course was split up into units of approximately nine lessons with each member of staff writing at least one unit per year group. To promote scientific skills and processes, problem solving has been adopted in a number of places throughout the course. However, here I will refer to the experience of planning, developing and using this technique in just one of the second year units.

One of my tasks in the development of this course was to write a second year unit on 'Pollution'. The aim of the unit was to develop an understanding of pollution and to encourage a social awareness of the problem and its possible solutions. Much of the unit is based on worksheet-led investigations and pupil research into books, videos and newspapers. Therefore, to increase the variety of activities and, hopefully, motivate girls and boys, it was decided that one of the lessons should be a problem-solving exercise. Through doing this exercise it was intended that pupils should:

1. Develop the practical skill of handling and assembling apparatus.
2. Develop the process of designing, testing and modifying an experimental approach.
3. Work together in groups.
4. Gain self-confidence in solving problems.

The problem to be presented to the pupils was 'Make this polluted river water fit for a fish to live in.' The problem was to be presented to eight science forms of quite different compositions:

- Three classes of 'selected' boys (i.e. had passed the Kent test at age 11).
- Five classes of 'unselected' girls and boys (i.e. had not passed the Kent test at age 11).

My own direct knowledge relates to one of the groups of selected boys. This class consists of 20 pupils who are brighter than average and who have been studying the 'Foundation Science Course' for approximately

1½ years. I have been teaching this group for nearly 1 year, and therefore know them well.

Their previous scientific experience included having performed many experiments, some problem solving, knowledge of separation techniques, how our water supplies are purified, and laboratory safety. Before presenting this lesson it was important to determine how the pupils might approach the problem. Other teachers wanted guidelines of what to expect and the laboratory technician needed to know in advance what would be required for this lesson. To address these matters I asked:

- If presented with this problem, how would I tackle it?
- Could I think of several alternative solutions?
- What was possible in 1 hour in a laboratory with fixed benches, gas and water supplies?
- What solutions were most likely to be suggested by the pupils?
- What apparatus were they likely to ask for?

To make the problem manageable by pupils, one teacher, and in 1 hour, I provided an apparatus kit for solving the problem. This also facilitated the technician's job and when presented to the pupils acted as valuable clues.

- Were other clues likely to be needed by the pupils?

I thought so, particularly by those less able to think things through for themselves, and it was important that this should be a positive experience for everyone.

- How could clues be given?

The teacher could monitor a group's progress and give prompts when needed or clue cards could be available. The latter method offsets the need for the teacher to be in more than one place at a time, and leads pupils away from considering the teacher as the only source of information. However, making up clue cards is time-consuming and I did not know precisely what clues would be needed at this stage. So I decided to leave this until I knew what type of help the pupils were likely to need.

- What size should the problem-solving groups be?

Educationally, the groups should be large enough to enhance the generation of ideas and develop discussion. However, they should not be so large that some individuals could sit back and let others do the work. Practically, there was a limit to the number of groups which could be equipped and serviced adequately by the teacher. Groups of three pupils were adopted.

- Should this lesson be trialled before introducing it to the second year pupils?

Normally the course was not trialled and the first participants were the guinea pigs. However, with a more open-ended lesson like this I thought it would be sensible to trial it first. I did this by presenting the problem to a third year group of wide ability. This had the advantage of seeing how a variety of pupils approached this problem. However, there were some drawbacks because the groups were older and had not experienced the new course, and the problem was no longer curriculum-dedicated. Nevertheless, two valuable pieces of information came from this trial:

1. More-able pupils wanted more apparatus (I had not thought of everything!).
2. Less-able pupils were confused by the amount of apparatus they were presented with, and for some the problem changed to one of 'how could all this apparatus be used?'

I attempted to resolve this dilemma by increasing the amount of apparatus available for the lesson, but initially only presenting the pupils with a portion of it. The remainder was held back in the preparation room and was only provided on request.

When the problem was actually presented to the second year class, they were told at the end of a lesson to give them time to think about it. They were shown the apparatus which would be available next lesson and asked to make a list of it.

On the day of the exercise pupils were asked to organize themselves into groups of three, to discuss each others' ideas, and to choose a method for tackling the problem. They were allowed 10 minutes to do this, and meanwhile the apparatus was arranged in sets around the laboratory. Each group could only use their designated apparatus and could not borrow from other groups. At the end of this planning session, I presented myself as a Safety Inspector, and all apparatus set-ups had to be passed by me before they could be used. General warnings about the wearing of goggles and heating sealed systems were given.

Pupils were told to have water samples available for examination by 10 minutes before the end of the lesson. While not encouraging pupils to seek help before thinking things through for themselves, I asked them to explain to me why they intended to use the apparatus in the way it was assembled. During these chats I did give them hints and tips.

Homework was set to describe (1) the method their group used and why they had chosen it, (2) to discuss what difficulties had arisen and how they had been overcome, and (3) what improvements they would make to their method if they were to do the job again.

During the follow-up session, a spokesperson from each group outlined their approach to the rest of the class. Opportunities to demonstrate their techniques and discuss improvements were welcomed by the boys.

A number of approaches were tried with a variety of apparatus, some simple and elegant, some resembling the alchemist's laboratory, but in all cases the pupils were enthusiastic in justifying their approach. The ensuing discussion showed there was no single best way of cleaning up the water, but several methods were possible. There was much debate on just how pure the water really needed to be! Questions were generated: how could their method be made more efficient, could any of the methods be scaled up?

The groups had tackled the problem in a variety of ways. Some had spent a long time carefully thinking about the possibilities before beginning. But most groups had preferred to get their hands on the apparatus at the first opportunity. Some learnt through having an idea, trying it out, and then modifying it. Others learnt from a less organized trial-and-error method, while some tended to copy but make their own modifications. This did not lead to friction because the other groups were too absorbed in what they were doing. It was interesting to see pupils applying to the problem their experience of boiling drinking water at camp.

The class had certainly enjoyed the exercise but had my original aims been achieved? The boys had assembled apparatus to their own design. They had had to think it through for themselves with no teacher demonstration or diagram to copy. Many more now knew about the hazards of heating a sealed system, when it was pointed out that their initial apparatus did not allow for the expansion of warmed air. They had appreciated the need for judicious positioning and tightening of clamps to prevent their apparatus collapsing. And if the skill of handling and assembling apparatus can be assessed by the amount of breakage, then they did well with only one beaker broken. The pupils had worked in groups; there had been arguments but of a constructive nature. Even individuals who had had disagreements in long-term projects had worked harmoniously in this shorter one. And, finally, every group had obtained some cleaner water!

CASE STUDY 13: SOLVING PROBLEMS – RELEVANCE AND CREATIVITY IN SCIENCE

Dave Wallwork

Much has been written about the use of problem solving. Indeed, many colleagues are convinced that they have been using the technique for years and cannot understand the sudden interest. However, it is my view that the closed technological problems to which they refer are a means of simply applying taught principles rather than the relevant, creative, more open-minded exercise that I perceive it to be.

I am convinced that the constructivist view of learning and its implications are very relevant to everyone. The technique of problem solving gives children the opportunity to work from existing mental frameworks, to modify these as appropriate and to build on them in a way which is meaningful to those involved. Problem solving is cross-curricular in nature and this idea is promoted if it is used, whenever possible, across normal timetable slots and involves teachers from diverse subject disciplines. Children also need connected periods of time to work through their problems. It is combinations of these factors which cause disruption in secondary schools and so hinder effective use of the technique.

With these ideas in mind, some colleagues and I organized a full-day problem-solving experience for 180 second year (12–13 year old) pupils of average ability and above. It also involved volunteer teachers from science, mathematics, English CDT and rural studies, together with technicians. All adults were involved in the planning, execution and evaluation of the day and very few had any previous experience of this teaching strategy.

The major aim was to examine the technique in action. The children were to be encouraged to solve problems in an organized manner along the Assessment of Performance Unit (APU, 1981) iterative loop guidelines and to record their feelings, fears, successes and failures along the way.

The planning was crucial and detailed criteria were used in selecting the problems to be offered. Each one had to:

- involve the application of several skills;
- be clearly structured and easily understandable;
- be interesting and relevant;
- have reasonable resource implications;
- be independent of detailed prior knowledge;
- be equally appealing to girls and boys;
- have a solution achievable in the time available.

It was also considered essential that there be a choice of problems so that the pupils themselves made decisions about relevance. The freedom of choice was extended to allow them to decide who they would work with for the day.

Pre-planning for successful problem solving would appear to be vital. When working, children need equipment quickly and, therefore, the problems need to be well researched by teachers in advance. Then a large pool of resources can be collected in anticipation of needs and so the discouraging effect of having to wait can be avoided.

The session began by giving the pupils a list of the problems and allowing them time for discussion among themselves before making choices. Each of the six problems offered was allocated a teaching space in order to concentrate resources. Here I focus on just one of the problems. It asked the pupils to design and build a device to allow a blind person to fill a mug safely with boiling water from a kettle. The numbers of pupils tackling each of the six problems was roughly equal – about 30 arrived to tackle the mug-filling device. The teacher introduced some safety rules and organized sub-groups to work independently on their own solution. His introduction involved all the children in the room, at the same time, in trying to fill a cup with cold water while blindfolded in order to experience some of the problem. However, the pupils were given no instructions or ideas as to how the problem with boiling water might be solved.

It was our observation that each small group of three or four worked very closely and tended not to be influenced by things going on around them. It was felt initially that groups would copy ideas and so stifle originality, but this was not the case. Communication within the groups was excellent and a shared understanding was quickly achieved. This was made obvious by the total involvement of two visiting French pupils, with only limited English.

The pupils seemed to respond to the relevance of the problem and worked tirelessly towards a solution. They worked through normal school breaks and there were no cases of misbehaviour. Plans were modified and in some cases changed radically in an attempt to produce a demonstrable solution before the end of the day. Initially the groups seemed to have natural leaders, but everyone gained in confidence and began to take equal responsibility. A great many valuable social skills were practised and this is one advantage of this technique. It was also heartening to see some girls who were known to be disaffected with science working in a meaningful way.

In their desire to produce working solutions, the recording of the process was very sketchy or non-existent. The idea had been for them to keep a record of plans and decisions, but the children tended to do this only when reminded. It was significant that failures were not recorded and that the emphasis on the role of written work in these situations needs careful consideration.

At the end of the day the children shared their solutions, demonstrating their devices with boiling water and blindfolds. Everyone produced a device and was given praise from peers and adults. No marking or comparisons were made by staff between individual devices and any evaluation was child-centred.

Comments from the children were sought in a follow-up English lesson when pupils were preparing materials for the school magazine. Some of these were considered important when the experiment was evaluated. These are typical of the view expressed:

The problem seemed impossible but with a little thought it could be done quite easily.

I enjoyed the day so much that I spent all but thirty minutes working. I should like to do this again.

Even though our group had thousands of arguments, I enjoyed it very much.

I think it is a good idea because you learn to work in a group and listen to other people's ideas. I think that is important.

It taught us that school is not just sit down, be quiet and do as you are told.

Children's reactions were also recorded at the end of the day which encouraged the staff involved. It was obvious that most had enjoyed the experience and it seemed that many had gained confidence in their practical abilities. Group skills had been practised and most groups had employed shared knowledge to effect solutions.

The question of what was learnt was discussed by the staff involved and the difficulty in measuring this and the whole question of assessing individual contributions was raised. This, like written work, seems to be an area for further consideration.

The difficulty of assessment is increased when the teacher is also having to be a facilitator, safety officer, sounding block and not just an observer. The staff involved found they had to resist the temptation to direct or intervene when pupils seem to be pursuing ideas which they felt were doomed to failure. Most staff found it very tiring and although they unanimously recognized its value as a teaching strategy it was thought to be something for occasional, not continual, use. However, despite the need for pupils to be prepared to make full use of this strategy and for a 'problem-solving teacher' to learn new skills, the trials carried out suggest strongly that this is one powerful teaching technique which should be used in schools.

CASE STUDY 14: PROBLEM SOLVING – FUELLING UP
FOR THE START

Mick Nott

You can say what you like about problem solving but it certainly makes you think. Some time ago we used to meet in our local SSCR working group and I found myself questioning the teaching methods and learning strategies I was using in my usual teaching. Admittedly the working group's focus was problem solving in *lower* secondary school science, and I had all age ranges to consider.

As much as we tried, we couldn't define a full list of criteria sufficient to *define* problem solving in school science. We decided that people should go away and try what they thought was problem solving and then report back to the group on their efforts. That seemed positive – at least it was doing something. Previously I had tried problem-solving work with classes as an appendage to particular work schemes. I thought these provided the background of facts, theories, processes and skills necessary to undertake the problems set. Now I was going to try to devise a problem which provided the opportunity to learn the content and processes for a much wider agenda.

In our working group we had discussed the role of the teacher when using problem solving and we agreed that the teacher had a role as a *guide* but not a *director*. The teacher could be authoritative but not authoritarian. The teacher also had a role as a problem setter: a person who would set problems which would provide a framework, or context, for purposeful activity and purposeful learning in science. The problems were to be much wider than those tied to specific parts of a course.

I saw classes twice a week for lessons of 70 minutes each. I saw third years only once a week, I shared my sixth form with a colleague, my fifth year were immersed in practical examination assessment, I had just tried some problem-solving work on heat transfer with my fourth years, and so, by process of elimination, I was left with my second years.

The work scheme for the group consisted of units of work that occupied approximately 6 weeks, i.e. about 12 lessons. The next unit was one on fuels. I was dissatisfied with it because, although the unit used discussion work and research, I thought that some of the theoretical content was inappropriate for the age range and that most of the practical work was either limited to teacher demonstrations with fossil fuel, or class practicals on foods. These require a high degree of manipulative skill, such as food tests. The whole unit was also questionable because the exercises were repeated in the fourth year work schemes for both biology and science.

Up to now the class had followed the other units in the second year work scheme and, although we had tried some problem-solving work, they were usually exercises that occurred as the last part of a piece of work. In the end, I decided that I would try some problem-solving work which would last for approximately 10–12 lessons on the general theme of 'Fuels'.

Starting points: I needed starting points! Approximately 4 weeks before the start of the work, I talked to a teacher who had changed some of the same unit by introducing (well-known) practical work on burning peanuts and introducing the calorific value of foods. The department had just acquired the 'Third World Science' materials and the next Sunday newspaper I read had a very moving, sensitive article on the uses of, and problems with, wood as a fuel in the developing world. That could be used to set a context. I read through the 'Third World Science' materials. Relevant to *fuels* there were the units on Charcoal, Distillation, Fermentation and Methane Digestors. There was a wealth of material and I had to decide what to use for resource material. I also had to decide on what I wanted the children to learn, and the way in which I wanted both the children and myself to work. I wasn't sure about assessing the work in the usual form of end-of-unit tests and recording the children's achievements as oral, practical and course work. An end-of-unit test would be difficult to write but I could still record children's progress in oral and practical work. I wanted the children to produce written work collectively as groups and so assessment of an individual's course work would probably be difficult. I was keen, though, to evaluate the unit and, in particular, see what the children felt about the work, the value they attached to it and what they thought they had learnt.

I decided too that the current 'Fuels' unit tended to concentrate on content and that our whole second year work scheme didn't give children enough opportunity to plan and devise their own experimental work. My aim became to provide children with the chance to devise their own experimental investigations. I wrote several objectives for the groups of children, such as the ability to plan a series of practical activities, ability to foresee apparatus requirements, ability to control appropriate variables, etc. And besides these 'abilities' I had a set of 'opportunities' in mind, such as the opportunity to consult written materials, to discuss ideas with each other, and to determine their own pace of work, among others. They were not all clearly articulated in this way, but at the time I didn't mind that. I think – just sometimes – that it's valid for a teacher to be 'goal-free' and just say 'that seems like a good thing to do. I wonder what they'll learn?'

I brought together lots of information and ideas about fuels, a few ideas on what I wanted children to try, but I recognized that I still had to organize something more firmly. I couldn't just say 'There's some fuels; find out about them.' I decided on four fuels: alcohol, charcoal, wood and sugar. Charcoal was easy to obtain as barbecue fuel; sugar was no problem and alcohol was readily available. Suitable wood for burning was hard to buy. All four are part of fuels commonly used in the developing world. My reasoning was as follows: firewood is a common fuel in the developing world; firewood becomes charcoal; sugar comes from sugar cane, and sugar can be changed to alcohol by fermentation. I came across a lot of information in the 'Fermentation' unit in the Third World Science materials about, for example, Brazil's use of alcohol to reduce their imports of petroleum.

I began by reading to the class a piece about the developing world's use of firewood and the kind of tasks that people (mainly women and children – many of their own age) had to do and how much time gathering this wood took up. From there I asked them about fuels they knew about, including gas and oil, and asked them to discuss and complete a worksheet in groups about the many uses of fuels. I asked them to see if they could cluster them into four types of use: transport, cooking, heating and lighting.

I had prepared some 'clue cards'. This was a sheet of paper or card which had some key questions, and/or statements, pictures, cartoons on them to set children thinking and give them some idea of what they could do. They weren't extensive, but brief and to the point. I had collected together as much reading material as I could from different sources (books, magazines, etc.), and I had a set of science magazines available with an index.

Their task was to decide through the design of their own experiments which of the four fuels available would be *best* for each of the four uses. The definition of 'best' was something each group would have to decide. The clue cards mentioned (in a roundabout way) things like temperature of flame, amount of light in flame, safety of storing the fuel, and so on.

Two restrictions were placed on the work. Alcohol always had to be burnt in spirit burners (all alcohol was kept, stoppered, in an adjacent prep-room and was only poured by staff), and any experiment had to be checked by me before starting. Other than that, children could ask for what apparatus they wanted.

It didn't begin well in the first lesson. There was an indifferent reaction to the article from the Sunday newspaper. The class couldn't seem to share my empathy with other people's hardships. They liked the idea of planning their own experiments to guide group discussions on the use of fuels and class discussions when sorting them into four groups. I explained what was expected of them, and that the spirit burner was for safety reasons. I indicated the boxes of books I had made available and they already had the clue cards available for the first bit of work. Everything seemed set for the next lessons.

From then on the unit worked extremely well. There was more than enough to keep them going in the early weeks, although I had to work harder towards the end to keep providing new background material. The last lesson was taken up in reporting on work and rounding off, to everyone's satisfaction.

SUMMARY

As in earlier chapters, there are four aspects to our summary:

1. Active learning;

2. Planning;
3. Organization;
4. What else is important?

1. Active learning

Mick Nott and Andy Howlett's contributions focus more on teachers than the pupils. That said, they give every indication that teachers who are new to problem solving emerge as very active learners themselves. The change of role required to cope with open-ended problem solving can mean a rapid reappraisal of self and function!

Bev Cussans and Dave Wallwork focus more on the youngsters and here there is every evidence that active learning is taking place. As Dave Wallwork says:

> The pupils seemed to respond to the relevance of the problem and worked tirelessly . . . through normal breaks . . . plans were modified and in some cases changed radically in an attempt to produce a demonstrable solution before the end of the day.

And, in almost a throwaway line, he says:

> At the end of the day the children shared their solutions, demonstrating their devices with boiling water and blindfolds. Everyone produced a device and was given praise from peers and adults.

Having been present on the day, we can say that Dave does scant justice to the exuberance and sheer delight that pupils derived from those demonstrations. They clearly felt they owned their own solutions – they had 'patented' an idea and made it real. What's more, they all worked!

2. Planning

As we said in Chapter 2, all practical work requires pre-planning. Some of the obvious aspects concern

- trying to guess what kind of special apparatus or materials will be needed;
- getting hold of a large amount of consumable items like string, sellotape, scissors, glue, cardboard, and so on;
- predicting what text and visual material will be needed to stimulate ideas;
- designing 'clue cards', safety instructions, pre-reading, homeworks, etc.;
- planning the timing, technical support, room layout, disposal of waste;
- planning the teacher's change of role.

In some cases the problem of apparatus was overcome by stipulating that the problem had to be solved 'using the equipment available' or with apparatus provided by the youngsters themselves. Beverley's approach was to almost 'over-order' and hold materials back in the wings so as not to overawe some of the groups.

3. Organization

Organization of time and physical space is important, particularly with problem solving. As in project work, youngsters may want to store work from one day to the next, and to have access to it in the interim. This needs flexibility and cooperation within the whole department, and the school. Apparatus may need to be ordered directly by youngsters and this needs a robust and responsive system that can cope with vagaries and omissions. As we suggested earlier, youngsters, like teachers, may be faced with the fact that being underprepared often means disappointment and the need to reorganize activities for the next session. One suggestion is to try to ensure that problems are 'do-able' in the time allotted, that the time allowed is generous (as in Dave Wallwork's case) or that there is a circus of problems (as with Andy Howlett's desert island theme).

4. What else is important?

First, a word about reporting results. At Ailwyn School, Dave Wallwork gave youngsters the opportunity to display their wares by demonstrating their device. There was sufficient variety in their approaches that the end-performance was entertaining as well as rewarding. He also asked another member of staff (a camera buff) to take photographs during the day and these were displayed as soon as they were printed. Posters are an obvious way to report outcomes and can be well designed to illustrate the processes that were undertaken and the results achieved.

The real reward for most of the youngsters was being able to design their own experiments and work in their own way – to own what they were doing. They helped to ensure that the problems were relevant and produced high levels of motivation.

BIBLIOGRAPHY

Adams, D. (1982). *Life, the Universe and Everything*. London: Pan.

Ashmore, A. D., Frazer, M. J. and Casey, R. J. (1979). Problem solving networks in chemistry. *Journal of Chemistry Education* **56** (6), 377–9.

Assessment of Performance Unit (1981). *Science in Schools: Age 13*, Report No. 1. London: HMSO.

Black, P. and Elliott, H. G. (1982). Problem solving by chemistry students. In *Qualitative Data Analysis for Educational Research* (Eds J. Bliss, M. Monk and J. Ogborn). London: Croom Helm.

Black, P. and Harrison, G. (1985). *In Place of Confusion: Technology and Science in the School Curriculum*. London and Nottingham: Nuffield-Chelsea Trust and NCST.

Cussans, B. and Munson, P. (1987). *Chaucer Foundation Science*. Canterbury: Kent County Council.

Further Education Unit (1986). *Supporting Science and Technology in CPVE*. London: FEU.

Howlett, A., Dalleywater, C., Nott, M., Clement, M. and Gibbons, G. (1986). *The Desert Island Problems*. Peterborough: Peterborough SSCR Group.

Kahney, H. (1986). *Problem Solving: A Cognitive Approach*. Milton Keynes: Open University Press.

Mathews, B., Scholar, J. and Hinton, K. (1980). *Problem Solving. Teachers Guide*. Mimeo. Roan School, Clissold Park School, London.

Munson, P. (1988). Some thoughts on problem solving. In *Problem Solving: A Collection of Ideas and Approaches* (Eds J. Heaney and D. M. Watts). London: Longmans for SCDC.

Newell, A. and Simon, H. A. (1972). *Human Problem Solving*. New Jersey: Prentice Hall.

5: ENCOURAGING AUTONOMOUS LEARNING

INTRODUCTION

It is an interesting paradox in education that we teach groups and yet we assess and examine the learning of individuals. This chapter brings together five case studies that are both similar and yet different. They are similar in that they focus upon teacher attempts to tailor classroom work to the needs of particular individuals, or small groups – to personalize their science instruction. Arguably this could be said of some of the approaches explored in other chapters, yet in these cases we feel that some extra effort has been made to individualize or personalize the work.

Classroom teachers may feel this is an esoteric exercise – they are seldom in the position to be able to spend time, energy or resources in shaping particular parts of the curriculum to the needs of specific class members. On the whole, groups have a range of topics to cover and this applies to all its members, with few exceptions. Teachers are also aware of the many problems concerned with trying to cater for a broad range of abilities within any one class without the added complication of trying to cope with matching individual dispositions.

There is a long tradition of thought and theory that lurks behind the work of advocates of personalized – or autonomized – learning. While we have no intention of undertaking an exhaustive review, we need to mention one or two interesting examples if only to try to place the contributions here into some context.

There are two main strands of work, which we refer to as 'personalized' and 'individualized' learning. The first has much to do with the traditions of Carl Rogers (1983). He talks about 'whole-person' learning, learning which is:

> pervasive . . . it is evaluated by the learner. She knows when it is meeting her need, whether it leads to what she wants to know,

whether it is illuminating the dark area of ignorance she is experiencing.

He continues by suggesting that the locus of evaluation lies within the learner, and that personal learning is of the essence:

> When we put together in one scheme such elements as a prescribed curriculum, similar assignments for all students, lecturing as almost the only mode of instruction, standard tests by which all students are externally evaluated, and instructor chosen grades as the measure of learning, then we can almost guarantee that meaningful learning will be at an almost minimum.

The second strand grows from the 'independent learning' tradition much in vogue in the late 1960s and early 1970s. Much of the early work in this field has been surveyed by Green (1976) and, though it may now seem to be slightly unfashionable, it is possible to see some of its modern antecedents in a range of current courses. Programmes of work, such as many of the Open University 'distance learning' methods, the Diploma in the Practice of Science Education (also a distance learning course) at the University of Surrey, or the Advanced Biology Alternative Learning (ABAL) project produced by the Inner London Education Authority (ILEA, 1985) are in use in schools.

Our five contributors tackle different parts of these two strands. Both Di Bentley and Mary Doherty focus on the needs of girls in tackling science. They do so in a very 'Rogerian' way by attempting to establish a climate of trust, moving towards participatory decision making, helping pupils prize themselves, build confidence and esteem. In Rogers' words, they encourage 'personal involvement' with learning.

Brian Taylor, too, discusses something similar. In his case he describes the PASS approach to DRIVE ways of working in a post-primary atmosphere. It is a primary science approach to be used in secondary schools which emphasizes the negotiation of contracts for classroom activities so that individuals and groups can pursue their own directions through a class theme, and 'work to their own ability and pace'. In this sense it borders on individualized learning, although he makes much of the cooperative atmosphere engendered in classrooms.

At the other end of the age range, and 'programmed activity' scale, Peter Richardson discusses the introduction of a fairly established independent learning scheme, the ILEA's ABAL scheme. His concerns are for the upper range of students – poorly motivated 'A' level students who arrive at lessons with very distinctive differences in academic background and achievement. He attempts to engender learning attitudes outside the confines of a recognized 'A' level syllabus. He focuses on the control each youngster has over the biology he or she is learning, and the learners' own evaluation of the task they have accomplished. The pace is set by individuals, or by a small group, and the motivation is all their own.

CASE STUDY 15: THE ABAL PROJECT

Peter Richardson

My school is a member of a sixth form consortium comprising four schools. The school shares upper- and lower-sixth 'A' level biology teaching with another school on the same site. There are two sets in both the upper- and lower-sixth which are comprised of students drawn from all four consortium schools. Each set is taught by teachers from both schools with two 55-minute periods in one school and three 55-minute periods in the other. The work scheme consists of syllabus topics together with practical and theory sheets developed between the two schools over the last 3 years. The syllabus has been divided and reorganized so that each set is taught different (but related) topics by their two teachers.

Few of this year's students are following traditional physics, chemistry and biology 'A' levels and thus prior knowledge in other sciences cannot be taken for granted.

Ten weeks after the start of the course (which had been used satisfactorily for a number of years) there was a general rise in concern of all four lower-sixth teachers involved. We expected a variation in achievement within the set but this year the problems seemed to be in the areas of general attitude, organization and motivation. There seemed to be a high level of absences and lateness to lessons. Course-work was returned late and was often of a poor standard and effort. Excuses and promises were, on the other hand, readily available and highly ingenious!

It seemed that, for most students, 'A' level study was a pastime: something to occupy them during the day. The 'new sixth' – often heralded – seemed to have arrived! This attitude was all the more alarming because of the efforts the consortium had made to counteract such feelings in the students. The lower-sixth starts with a 2-week subject-specific study skills course along with school provision of general study skills support and a departmental offer of individual biology support for three periods a week. It was becoming clear that our traditional method of practical, lecture and private study (both general and specific set work) was no longer effective.

The aims of change

It was decided that we must provide a much more structured course so that learning activities (lessons) and private study assignments (homeworks) could be presented in a way to aid students in their planning and organizational skills. Our aim was that the success in knowing exactly what needed to be done, and keeping to deadlines, would improve motivation and effort. We did assume a reasonable level of intrinsic ability and motivation to complete the course and pass the exam.

The restructured course – reasons for choice and expectations

Rather than spending considerable time and effort in rewriting and reorganizing our work scheme and associated materials we decided to trial an individualized scheme already developed. The Advanced Biology Alternative Learning (ABAL) Project (ILEA, 1985) materials had been written for exactly the situation that our schools were experiencing. The materials consist of 10 students' guides that integrate theory, practical, and teacher and student self-assessment activities. There are also associated audio-visual materials and the teachers' guides contain much useful help and hints for both teaching biology and running an individualized learning programme.

It was decided to introduce the materials to the current lower-sixth groups and run the rest of the course based on the ABAL scheme with the intention of continuing into the upper-sixth if the trial showed signs of improving performance and motivation.

There were two main reasons for our choice. First, we were familiar with the materials since both schools had done some of the trialling of development work. Second, were the written materials themselves. The format of the student guides would help provide a consistent approach to study between the two schools and the four teachers. The nature of the study guides, where about 9 weeks work is presented with objectives, questions, summary assignments, etc., allows a more structured and coherent programme to be presented to students. For students absent, work missed is immediately available to them in a form that they can use effectively to catch up without needing massive amounts of teacher time to cover basic ideas and information.

The project makes the premise that teacher–student contact time is too valuable for the essentially routine activity of transmitting information by lecturing, since such methods do little to help students learn actively. Such contact time should be used for more stimulating and motivating activities which help students to interpret, understand and use scientific ideas. The most important resource in the classroom (the teacher) is now free to stimulate learning by direct contact with the students, and thus is able to challenge their ideas. The integration of self-assessment activities into the student guides helps develop active (reflective) reading and a more critical awareness of their own understanding and give immediate feedback to students on their own learning and progress. Students can then formulate their own questions rather than just respond to questioning from the teacher, which may or may not have relevance to their own conceptual difficulties. This 'guided' nature of the texts allows students to consider fairly complex ideas or situations without the teacher needing to be on immediate call. Able and/or organized students can preview work coming up, those with some difficulty can spend extra time reviewing the work they found difficult. These activities can be done either in class time or during private study: true independent and individualized learning can take place. After the students' initial structured encounter with concepts, the skills of the teacher can then be better used to stimulate and help students reconstruct and consolidate their ideas. With the careful presentation of more open-ended and less structured work, students could gradually be given more responsibility for their own learning at a time when they are more likely to encounter success with the tasks set.

Pre-planning

Once the ABAL materials had been seen as a possible answer to our problems I canvassed student opinion. I explained that the different format was to help them study more effectively and I outlined some of the points of why I had chosen this particular scheme. Their comments indicated that they were not totally hostile to the materials and methods. Although it would take much longer than this to familiarize the students, they were versatile and could adapt. The biggest problem would be to change teacher attitudes and practices.

I then raised the idea of using ABAL with other science teachers. One important consideration was that the other teachers were happy with the information in the student guides and the way it was presented. It was also important for teachers to realize that although it is desirable to use the materials as presented for the first time of operating, they were free to substitute other activities where they thought fit. Once the other teachers were prepared to trial the materials, the technical support was considered. The ABAL materials were developed with standard equipment and practicals in mind and little 'special' apparatus or methods are needed. For each 9-week student guide the approximately 10 practical exercises did not present a significant increase in technician workload or demand for apparatus. However, all of the reasons for change and how our teaching would be different was explained to the technicians. The teachers' guides give very clear apparatus lists, technical notes, etc.

Bringing about change in the way students and teachers approach and work with ideas is never easy. I found that careful planning in advance helped to give a sense of support and continuity so that problems were confronted almost before they became problems.

The teachers' guides already give much information on the pre-planning needed to operate the learning system in general. The most important part of the planning I needed to institute was that time be given to becoming familiar with the learning activities and the scheme of work. Once familiar, a timetable for laboratory work (both practical and theory), private study assignments and completion deadlines could be worked out. An overall 9-week unit schedule and a specific 3-week assignment work plan is needed. I also found it helpful to annotate my own copy of the students' guide in pencil with the dates on which various activities were done and work set, etc. This helps in planning and also shows students who have been absent exactly what was done and when.

The final aspect of the pre-planning was to gain some sort of objective measure of the impact of the new materials and methods on the students and published inventories helped here. For those new to this type of work, lesson organization should involve a brief introductory lecture which precedes group practicals in laboratory time. A section of text with its related assignments is set for homework as individual study. The next lesson looks at the work covered either by going over expected answers or by application or illustration of the concepts in a new situation not given in the texts. If any problems of understanding are evident in some students, they can be given a short tutorial session while the rest of the set continue with the next block of text. Formal assessment occurs every couple of weeks with either self-tests or past exam questions. At the start of an individualized scheme, whole-group teaching paced by the teacher should be the dominant format allowing students to become familiar with the materials and keeping to deadlines.

With the use of the materials, and as the students become more confident and responsible for their own learning, the individual study time can be extended to a couple of weeks. Dates when practical apparatus will be available need also to be given so that students can plan their own sequence of study. When this point is reached, lesson time can be used much more for individual or small-group tutorials where specific pieces of work, concepts or difficulties can be considered.

Evaluation

At present we are only half way through the trial period and the work is still in the rather structured phase which makes it difficult to see changes in motivation and student self-organization. However, some aspects have become clear already. Students have a better knowledge of what is expected of them and what they should know and be able to do at the end of a unit of work. They have learnt the value of using texts of many kinds to explore scientific ideas and look for evidence.

CASE STUDY 16: TEACHING PHYSICS IN AN ALL GIRLS SCHOOL AND WAYS OF PERSONALIZING THEIR LEARNING

Mary Doherty

This paper concerns ways in which changes can be, and have been, put into practice so that the science curriculum could be developed for the particular needs of girls (see also Doherty, 1987).

The school is a four-form entry 11–16 school for girls, situated in an area where there is considerable demand for single-sex schooling. The school evolved from an 11 to 14 high school and we are currently in our third year of preparing pupils for public examinations. We set about encouraging girls to study the physical sciences and in doing so highlighted areas of concern we needed to address. One area giving concern was the considerable dependence of pupils upon teachers for reassurance and support. This was particularly worrying lest the level of support could not always be extended. The first round of pupils studying physics encountered considerable difficulty because they did not have any older pupils either as role models or as sources of support.

When the girls started their physics course in the fourth year, care was taken to make the transition as gradual as possible. During the summer holidays a day was spent in discussion with the head of the maths department about the type of mathematical problems the girls would encounter, the timing of the introduction of certain topics, agreed terminology and the layout of problems. This discussion created the basis for close liaison between the two departments throughout the fourth year.

For the remainder of the staff, I gave a short talk concerning the reasons why girls find physical sciences difficult. I explained what the department was aiming to do in helping girls succeed in physics and highlighted the notion that the first term of any such venture would prove the most arduous. I asked for, and got, considerable encouragement and support from form and year tutors in encouraging the girls in their efforts.

Girls soon began to have problems coping with the demands of homework, not just in science but in all subjects, exacerbated perhaps by the inclusion of both geography and history as compulsory subjects. We negotiated with students and agreed that homework would be set on Friday to be handed in the following Wednesday. I agreed to be available at break and lunchtime on Mondays and Tuesdays to help with any difficulties. Throughout the year it became increasingly apparent that girls were committed and motivated to succeed in physics, chemistry and biology.

There was still a major problem. The girls seemed all still to be thinking of careers in hairdressing, nursing, banking and 'working with animals'. My preference would be that they choose *not* to be scientists deliberately, basing their choice on good information than do so from ignorance or a perceived lack of choice. Videos and films produced for WISE year were borrowed and shown intermittently to the fourth year. Mary Ayre of the Polytechnic of the South Bank was invited with two of her female engineering students to discuss their studies and training with the girls. They were then invited back for a second visit by very popular demand. Two male engineers from Opening Windows on Engineering (OWOE) also visited to discuss projects they worked on. One, a chemical engineer with Unilever, described 'making a silo' and involved his audience in re-enacting the project and asking for suggestions for overcoming various problems. Similarly, the second – from British Rail – generated a ferocious debate about electrification on the Tonbridge to Hastings line, and about the exact locating of a sub-station. During the year we also took 60 girls to Engineering and Industry Training Board's (EITB) open day.

We maintained progress throughout the year by means of tests and exams. These showed that some girls were coping much better than others. Each girl was interviewed after the tests on an individual basis to explore which aspects of the work she found difficult. Since the timetable gave me only one lesson a week for 2 hours, this meant starting each new lesson almost 'from scratch'. I decided to run a mid-week 'clinic' after school to help with any problems. Attendance was entirely voluntary. On the first evening almost the whole 'O' level group attended. Some did not actually need help, they really wanted reassurance, while some needed a little – and some a lot – of assistance. I failed to discourage those who were coping from attending, they pleaded to continue. Therefore, I eventually split the groups into three and ran three after-school 'clinics' instead. At the point when there were both fourth year and fifth year physics students it was impossible to offer both years the same level of support. So I asked each fifth year to adopt a fourth year.

Early in the autumn term of 1985, a party was organized during the lunch hour, and the fourth year physicists were introduced to the fifth year physicists. Prior to the party I had explained to both groups that I wished the fifth year pupils to 'adopt' one or two of the fourth year pupils to help them when they encountered difficulties with their work. The fifth year immediately empathized with the fourth year and were very willing to help. Once the ice had been broken at the party many girls arranged their own adoptions, leaving me to arrange only a few. I agreed to make the physics lab available at breaktimes and lunchtimes and the fourth year pupils made their own appointments at the fifth years' convenience. I had already explained to the fourth year that I would give them five nights for homework so it was their responsibility to attempt the homework on the first night and to make any subsequent arrangements with their own fifth year helper as soon as possible. The fourth year pupils did not need fifth year help throughout the year. Some exceptions were when we were studying such things as the gas laws, specific heat capacity or total internal reflection: despite collaborative efforts on the part of the Maths and Science Departments their capabilities in maths did not match the demands of the physics syllabus. The fifth year pupils thoroughly enjoyed the experience and it gave them valuable opportunity for clarifying their own ideas and for revision. The present fifth year (the adoptees), therefore, have not developed as great a dependency on staff and are, on the whole, coping very well. They show much greater confidence than the previous group and rely on each other for help and support. There has, as yet, been no need for the current fourth year to be adopted, despite repeated requests from the fifth year to be allowed to adopt a fourth year!

CASE STUDY 17: SCIENCE WITH A SOUL – LEARNING FROM FRIENDS

Di Bentley

For several years now, the 'problem' of few girls wanting to continue with science as they get older has been focused upon in writings on science education. Alison Kelly (1987) has made the point more recently that to construe the problem as being one of enticing girls into science may not be the most productive way of progressing. Rather, reconstructing science and in particular science education so that both are more in keeping with the experiences and explanations of the world that are familiar to women, might serve teachers better. The writings of Sue Rosser (1986) give some ideas as to how this might be brought about. She has suggested five goals for interactions in science classrooms which she describes as 'feminist pedagogical methods' or, in other words, using ways of managing learning that reflect the social experiences of women. Her five approaches are:

• establishing an atmosphere of mutual respect, trust and community;
• sharing leadership;
• organizing cooperative structures;
• integrating feelings and ideas in science;
• action.

This case study looks at how some, if not quite all, of these might be achieved in science teaching.

The work described here was conducted with third year pupils in the first instance, but has since been used by fourth year GCSE groups. It took place as a piece of collaborative teaching between myself and a teacher colleague who was trying out new teaching approaches. Its basic premise is that in a class of 25 pupils and one teacher, there are 26 experts, and 26 people capable of directing, organizing and facilitating learning. The topic area is irrelevant – in this particular case, the pupils were working through a unit on materials.

The original intention was not to set up a set of conditions which would match Rosser's goals. The discovery that the conditions did so rather neatly, came from reading very much *post hoc!* Rather, it was to assist pupils in learning from their peers and taking responsibility for major parts of their own work. I was responsible for setting up the process; therefore, prior to the lesson, at the beginning of the unit of work, pupils had been divided into groups with five pupils in each group. I had also chosen a leader for each group. It would have been perfectly possible to allow the pupils to elect their own leader, but as I did not know the pupils well, I was uncertain of what self-selection might hold in store. Each group leader was given a guide pack for running a group. These were a set of simple suggestions on managing groups, e.g. hints outlining how disagreements might be dealt with in the group, how to summarize ideas periodically if discussion had been ranging fairly wide, some techniques for ensuring that everyone got the opportunity to contribute both ideas and actions and a few ideas for helping people to assume particular responsibilities in a group task. While the rest of the class were set a homework designed to make them think about the sort of tasks they usually performed in their science working group and how they con-tributed to the success of a science experiment, the group leaders were required to read their pack. During the next lesson they were asked to explain briefly to their group how they, as group leader, thought they might conduct group exercises and what particular responsibilities the leader and group members might have to fulfil. The groups were given the opportunity to make further suggestions to their leaders and then the real task of leading the groups through the work began.

The focus of the unit of work was materials; the particular aspect that the class was studying was fabric technology. The whole class was set the task of finding the most suitable fabric for clothes to be used under the following conditions:

- for making ski jackets;
- to protect sailors from the waves on board north Atlantic trawlers;
- as material for children's nightclothes;
- as coats for lollipop persons.

The team leaders were asked to convene a meeting of the whole class to agree what the main areas of investigation should be and which team should do which parts. After some difficult initial discussions, chaired by the team leaders, different teams eventually agreed their contributions to the whole task. The investigations were to be:

- the degree of fireproofing of the materials;
- their insulating properties;
- their capacity to withstand wear;
- their ability to take strain;
- their capacity to take up dyes;
- the waterproofing of the material.

The groups negotiated their time-scale for each of the investigations and the leaders set about organizing their own teams. It was agreed that at the end of the work, the whole class would prepare a marketing and advertising brief for the materials which they had found to be best for particular conditions. The team leaders were required to direct the planning of the operation, ensure that everyone undertook appropriate activities best suited to their skills, kept to the time allocations (three double lessons) and contributed to the overall report.

Almost all the team leaders went about their jobs in different ways. One boy allocated tasks to the team members, organized what should be done and spent his time carefully directing and coordinating the work of other people. This team obtained accurate results and got the job done very efficiently but, on the whole, there was very little cooperation between individuals in the team. Each conducted their own part of the task and wrote up their own report.

Another group tackled their work on flameproofing by planning the whole investigation together and setting up a sort of production line – one person to make different concentrations of flameproofing solution, another to cut the materials to size, a third to immerse them for the required time, a fourth to burn them and the fifth to record results.

The most interesting group – a mixed gender group – negotiated the best way of doing the experiment, each team member being invited by the leader, in turn, to suggest a way of (in this case) finding out which material would take the most wear. They agreed on two possible methods of doing the experiment and the team divided itself into two pairs, with the fifth person being a 'tester'. They set themselves one lesson (half of a double) to run a pilot of each method, with the tester going round to decide which was the easiest or most productive method. In this way they alighted on one investigative method for the whole team to pursue. They then allocated themselves different parts of the task and worked out a questionnaire to find out what consumers thought were the most hard-wearing materials, which they all duly conducted for their homework.

Every step of their investigation was done by careful cooperation. They reconvened as a whole group regularly, while working on practical tasks in pairs during the next lesson, to report on progress and share new ideas. Finally, they wrote up their report as a collaborative effort, drafting and agreeing the text between them.

The final session, after all the groups had finished their reports, was a time to share findings and sort out the marketing of the various 'most successful' materials in the trials. It had the recipe for instant chaos! There were five groups, all wanting to report their findings, their innovative approaches and their excitement – well in most cases! There was, in short, a tremendous amount of work to be got through in one lesson, because after the reports, the work of planning the marketing had to be done. With hindsight, I am not sure that I would do it that way again. I think two sessions, one for report and one for planning, would have been better, and the group leaders were quick to point this out. They shared their concerns 2 days before the final lesson, and spent a great deal of time together at lunch and breaktimes, planning the organization of the report-back, each leader then briefing their group as a result of the planning sessions. It was due to their careful organization that the work got anywhere near completion. In fact, I think it would have been impossible for one person – even the teacher – to have achieved the same thing.

In the final lesson, then, the group came together as a whole class, listened to each other's brief reports, which had also been displayed for easier reading later, and began to discuss the means by which they might advertise and market the materials for their various properties. One group, via their leader, suggested that they should describe a day in the life of a lollipop person before and after having a waterproof coat made of the material they had tested. Other suggestions were just as interesting and wide ranging, from the 'product profile' scientific report of the efficient group, to the drawings and cartoons of another. In the end the product profile idea was adopted by the whole class, supported by the drawings of another group, and everyone worked at converting their findings into the agreed format.

Over the whole set of lessons, the majority of the group leaders kept their groups to task, ensured responsibilities were fulfilled, and made themselves responsible for keeping the work together. One was more reluctant than the others but he was pushed along, albeit somewhat unwillingly by the others. As well as meeting for planning the final lesson, some of them met occasionally at breaktime before the lesson to see how things were progressing and how they might manage the next part of the work. One leader suggested to some of her members that since their work was almost complete, they might help another group who were behind schedule.

On reading Sue Rosser's ideas somewhat later, I wondered if our method of 'learning from friends' had been a way of developing some of the 'feminist pedagogical approaches' she outlined. The answer, I suppose, is both yes and no. Yes there had been cooperative structures within individual groups, and yes leadership had been shared – if only between the teacher and five others. One group had managed to integrate feelings, in terms of the lollipop person, into their ideas and results of experimentation. Certainly, we had established an atmosphere of mutual respect for the work of other groups, some trust between groups and, for most of the group leaders, some idea of community. Individual members of groups had been able to appreciate the skills of one or two group members who had worked hard at being good negotiators and careful listeners.

What role had the teacher and I played? We had had the peculiar luxury of being able to ask real questions of pupils that made them begin to speculate about the scientific ideas and principles that lay behind their findings, questions that made them rethink their investigative approach, re-examine their findings and prepare to re-do bits to produce better results. Of course all this could only occur once we had got over the first rush of ensuring that the equipment that was ordered was suitable. We enjoyed sharing the responsibility for the lessons with pupils and were happy to take suggestions from group members and leaders for revised procedures for the next lesson, or next time the activity was done.

As for the pupils, most of them clearly enjoyed it, not least because it was novel and they had a hand in reshaping what happened. Not all of them of course wanted to be group leaders or enjoyed the experience. Some resented their group leader and wanted to refer to the 'teachers' all the time, in order to override their authority. This was obviously something we needed to watch out for carefully. On the whole, though, I would say that it was a reasonable success.

Speaking personally, and again with a great deal of hindsight, I would hope that this approach moves some way towards the goals of Sue Rosser and brings us a step nearer the aims expressed by Rita Arditti (1976) when she said:

> The task that seems of primary importance . . . is to convert science from what it is today, a social institution with a conservative function and defensive stand, into a liberating and healthy activity: science with a soul which would respect and love its objects of study and stress harmony and communication . . . When science fulfills its potential and becomes a tool for human liberation we will not have to worry about women 'fitting in' because we will probably be at the forefront of that 'new' science.

CASE STUDY 18: A PRIMARY APPROACH TO SECONDARY SCIENCE

Brian Taylor

This case study reports the work of a group of Essex teachers (1988) who as part of their work for the Secondary Science Curriculum Review decided to examine the problems experienced by pupils moving from the primary to secondary phases of education.

The working group looked at specific aspects of primary science that obviously encouraged the pupils to investigate a variety of phenomena associated with the current topic of study. It quickly became clear that the primary school offered a different approach to the learning of science than that of the secondary phase. This decided the focus of the work: producing an approach or way of working that could be adopted by secondary school science teachers which would enable a less traumatic transition for most primary school children. The basic idea and philosophy behind this way of working became known as PASS.

PASS is not a new lower-school science course. Rather it is a novel way of approaching the learning requirements of children, especially in the first and second years of the secondary school. It takes into account good scientific practice drawn from our observations in primary schools and puts these in the context of the deeper involvement of pupils in developing science concepts and skills.

The main divergence from traditional secondary science work is the emphasis not on knowledge or concepts but on the DRIVE method of working. This is really using the natural innate ability of the child to ask questions or pose problems and then formulate an answer or seek a solution. It may be summarized as follows:

Curiosity, problem, question	$\longrightarrow$	Idea	$\longrightarrow$	Design of test, activity, investigation	$\longrightarrow$	Carry out test, activity, investigation

The DRIVE approach enables all pupils to operate scientifically, experience success and a degree of personal fulfilment whatever their knowledge, concept and skill level.

Because this approach is designed as a way of working, it does not focus on content areas or knowledge. Teachers can use it to teach many different concepts and skills which they want to explore with their class. What aspects pupils choose to learn about is likely to depend on their previous knowledge. Thus it is quite feasible that within the class there will be a large number of investigations and experiences being tackled at the same time. The approach encourages both teacher and pupils to become involved in and explore a range of study techniques by using relevant, up-to-date and appropriate information about their aspect of science. This helps to create the real opportunity for science to breach the traditional subject boundaries found in the secondary school.

The approach is seen as a foundation for science education and it is suggested that it lends itself to the 'spiral of learning' conducive to problem solving and technological investigations.

The approach involves the teacher and pupils formulating and agreeing a 'contract' as to the scope and nature of the investigation. The teacher can dictate general topic areas, but then decides, with the class, the various activities and investigations to be carried out. This is clearly a continuation of the topic or theme approach adopted by many primary teachers. It relies on a high degree of teacher organization and skill, i.e. the ability to keep several balls in the air at once and remember whether each one is going up or coming down! Secondary teachers may well learn a great deal from primary colleagues. For example, work should be initiated by a stimulating 'starting point' which should, ideally, encourage most, if not all, the pupils within a class to develop their own ideas for investigation.

The organization of the class may also be seen in a more fluid aspect, since the approach involves a variety of teaching styles. While pupils are conducting their open-ended investigations, teachers may be counsellors, helping to provide simulations, assisting in problem solving, showing how to use video-recorders, or seek information from a computer program. Some pupils may work together out of common interest, or common area/depth of study, whereas others may remain in friendship groups or even work individually.

It was thought most desirable that the area of work a pupil was engaged in was seen by that pupil as emanating from within, i.e. having relevance to the pupil. The basic premise of the DRIVE approach is that of internalization of the problem to be investigated. Thus the work carried out by the pupil should originate from the pupil and not the teacher. The teacher, therefore, takes on the role of a resource and facilitator rather than provider.

The PASS approach allows great flexibility and encourages the development of various communication skills while giving considerable opportunity for self-directed study.

Exemplar materials have now been developed around five main themes: metals, tree leaves, time, litter and soil. The materials were trialled in 12 secondary schools and a special school for the physically handicapped. The materials were used with pupils covering the complete range of ability. Each participating school was visited and the teachers were encouraged to use the full potential of the PASS approach. Many lessons were delivered on a team teaching basis between the class teacher and a visiting working group member.

When a topic had been completed, all staff involved were requested to complete a simple evaluation form. They were expected to comment upon the work carried out by the pupils, the starting points used, the way pupils worked, and so on. Most teachers also wanted to assess their pupils to determine whether or not they had gained from the DRIVE experience. The working group developed assessment schemes for each topic produced. These schemes were given to participating teachers but each teacher was allowed the freedom to use them or not as they wished.

The school trials and teacher evaluation provided some valuable information. The use of the DRIVE approach tended to allow pupils to work to their own particular ability and pace. The vast majority of pupils reacted very favourably to the freedom of choosing their own area of study and method of investigation, although some needed more careful watching and occasional pushing.

The work required for preparation and planning was greater than with more straightforward science teaching approaches but the benefits were seen in a greater enthusiasm among the pupils and in many cases a dramatic improvement in their work. This subsequently led to many teachers experiencing greater fulfilment and job satisfaction. The atmosphere within the class, although busy, and sometimes fraught, was also filled with a feeling of great enjoyment and relaxation. People were working hard, yet in a relaxed way.

Most pupils enjoyed working in this way. A small survey of pupils in one particular school showed that pupils were highly in favour of this way of working. Teachers reported that there was a decline in enthusiasm when they reverted to the more formal way of working.

The administration of lessons where pupils were at liberty to choose their own apparatus caused some concern at first. Most teachers adopted a 'booking' system for apparatus where pupils were required to notify the teacher in advance of the apparatus required. They soon became accustomed to this ruling and lesson preparation and planning became as much a part of the pupils' responsibility as the teacher's.

Monitoring of progress is more difficult. The approach puts forward a strong case for continuous pupil profiles to be used, because not all pupils are engaged in the same work at the same time. It is important that a record of work completed is kept for each child in the class. This record can obviously be passed on to the teacher for the following year, thus clearly indicating the exact areas of science experience each child has covered.

One great disadvantage for this approach is the structure of the secondary school day. Most teachers commented on the fact that pupils were very often frustrated by the bell ringing just as they were 'getting somewhere' in their investigation. There is obviously a great deal of thought and discussion necessary between teachers and school administrators if this method of teaching is to be adopted. The reactions from most teachers and nearly all pupils is that any effort put into facilitating this method of study would be more than worthwhile.

The evaluation comments received from teachers outside the original group have served to confirm the belief that science study for secondary school pupils can be made more stimulating and worthwhile, while still providing the 'in-depth' knowledge most secondary teachers seem to believe essential for secondary study. The essential ingredients seem to be the need for the teacher to accept the position of resource rather than deliverer.

With the move to balanced science in the examination years it would seem realistic to continue to teach the young secondary children in a way they have already accepted as worthwhile in their primary school. We, the secondary teachers, need to look at our responsibilities for providing an atmosphere and curriculum suitable for pupil needs. The situation is clear: when children enjoy their work and are personally involved in what they are doing, when they are studying an area that has arisen from their own interest, then they learn, and they retain what they have done far better.

SUMMARY

To summarize, we consider our usual four questions:

1. Which aspects of active learning do they encourage?

In case study 15 Peter Richardson makes a telling point:

> . . . teacher/student contact time is too valuable for the essentially routine activity of transmitting information by lecturing, since such methods do little to help students learn actively.

This seems to summarize the import of each of the case studies. Prising the teacher out of the purely transmission mode allows greater freedom for them and the students. There is a place for transmitting information, as is obvious from what is said. But the transmission is geared to establishing the context for the youngsters to negotiate their own avenues of approach. Teacher-as-lecturer is at a minimum and teacher-as-consultant is maximized.

2 and 3. What planning is required and what organization is needed?

In many ways there is not too much more to add here. While the case studies do not actually provide checklists for what needs to be done in advance, they are fairly clear and unequivocal. It is obviously a lot of hard work, whether you are organizing a party for fifth-years to adopt fourth-year physicists or persuading teachers and students to adopt an 'off-the-shelf' independent learning package.

4. What else is important?

The considerable 'lead in' time is important, because even relatively minor changes to current practice and routine require time to come into being. Some of this time is necessary for adequate consultation: as Peter Richardson points out, students were canvassed, teachers were introduced to the materials, technicians were to be involved in the discussions. Or in Brian Taylor's case, work by primary teachers was to act as the catalyst for secondary school work, a pleasing inversion of the more usual hierarchical 'top-down' imposition of curriculum development. Needless to say, consultation is time-consuming, and sometimes counter-productive. In case studies 16 and 18 the students seemed to be very enthusiastic about the changes; this is less clear cut in case study 15. Peter Richardson gives the impression that the start of the project seemed to be making progress but that it was yet early days to make too many bold statements.

BIBLIOGRAPHY

Arditti, R. (1976). Women in science: women drink water while men drink wine. *Science for the People*. March.

Doherty, M. (1987). Science education for girls: a case study. *School Science Review* **69** (246), 28–33.

Gilbert, J. K. and Horscroft, D. (1985). Diploma in the Practice of Science Education (DPSE) Module 0: Introduction to the Course. University of Surrey/Roehampton Institute.

Green, E. (1976). *Towards Independent Learning in Science*. St Albans: Hart-Davies Educational.

Inner London Education Authority (1985). *Advanced Biology Alternative Learning Project*. London: ILEA.

Johnson, J., Taylor, B. and Soar, G. (1988). *Primary Approaches to Secondary Science*. Chelmsford: Essex County Council.

Kelly, A. (1987). *Science for Girls*. Milton Keynes: Open University Press.
Rogers, C. (1983). *Freedom to Learn for the 80's*. London: Charles E. Merrill.
Rosser, S. (1986). *Teaching Science and Health from a Feminist Perspective*. New York: Pergamon Press.

6: GAMES AND SIMULATIONS: AIDS TO UNDERSTANDING SCIENCE

INTRODUCTION

Games and simulations are not new in the world of education. The use of a game to reflect the problems of the real world goes back as far as at least 3000 B.C. to the Chinese. Gaming might best be defined as a structured system of competitive play which incorporates the material which is to be learnt. Simulations are a little different in that they are less structured and not competitive. They aim to present the student with a highly simplified reproduction of either a real or imaginary world. They may be cardboard models, letters, telephone messages, or be based around the actions of large social organizations such as the members of an industry and the town officials and residents with whom they come into contact. Many simulations involve decision making – such as Margaret Davies' 'Cherry pie game' – or communicating and negotiating with each other. The students are the human element in the simulation or 'system' and they are expected to react to the situation in a way that will be determined by how they and others see their role within the system.

Games, such as card games, board and dice games, offer according to Brandes and Phillips (1979), 'the promotion of effective communication'. Brandes and Phillips find games very useful. They state that 'by helping people to relax in groups, games can promote the flow of communication between complete strangers and particularly shy people who need encouragement'. Thus games are an easy way of facilitating the exchange of ideas which are involved in many of the aspects of active learning in science. The use of games during the exploration of scientific ideas can help to extend or clarify particular aspects, reinforce previous ideas, or introduce new ones through analogy. They can also help to provide some of the social and technological contexts which are an important part of science education, yet often difficult to introduce in a meaningful way.

More importantly, games and simulations can make learning about science fun. They can act as a mechanism by which groups can work together, develop cohesion and enjoy the whole experience.

Jones (1985) believes that games, and in particular simulations, have a special contribution to make to education:

> . . . [This is because] . . . they are characterised by action and behaviour rather than by the acquisition of facts. They are about verbs rather than nouns. The students become active and powerful participants in the learning process and this has a very profound effect on both students and teachers.

Margaret Davies, in her case study, sees four aims for the use of both games and simulations in science. She states these as being:

- to take science outside the laboratory;
- to take a broader view of science than that provided by laboratory experimentation;
- to show relevant examples of science;
- to make students think.

Ken Jones (1985) draws a careful distinction between games and simulations. He makes the point that games are competitive, whereas many simulations are cooperative, not competitive. Further, he claims, it is the actions of the participants in games and simulations which are different. The object of a game is to win and the actions of the players are geared towards that end. In simulations, even if they are competitive – wanting to build the factory or prevent it being built – they are person-to-person activities, not individual gain activities.

Rice (1981) takes the view that simulations are important because they involve problem solving. They can, he claims, 'permit us to explore various courses of action without having to actually bear the costs of a wrong or dangerous decision'. This is certainly the case with the kind of computer simulation which Anita Pride describes. In her computer simulation, many of the decisions taken to bring about a solution to the problem are potentially fatal for the individuals involved – a powerful learning experience!

Linda Scott's contribution involves a card game, used to give the players confidence in areas of physics. This clearly has the hallmarks of a game not a simulation, in that there are rules by which it is played and there are some elements of competition. There are also, from Linda's description, a great many areas of discussion and cooperation as well, especially in the area of designing new games with the cards. Norma White also uses card games as a way of making students think about the issues involved and providing time for sharing ideas and discussion around a particular focus.

CASE STUDY 19: PLAYING GAMES IN PHYSICS

Linda Scott

There is a lot of evidence to suggest that, in general, the hobbies and pastimes of boys outside school gives them certain initial advantages over girls when studying particular science topics, notably in the areas of physical science (APU, 1985).

The technique described below was devised in order to approach one such area of the curriculum. The student activities were planned to allow individuals to work at their own pace and at the level appropriate to their previous experience. In this way any youngsters who were less familiar with aspects of the topic initially were able to develop their own ideas and understanding in a non-threatening environment.

The activities were used as an introduction to the topic of electric motors with third year students but similar work could be used in other curriculum areas. Previous experience in presenting this topic to mixed gender classes had indicated that many of the boys were much more aware of the diversity of uses of electric motors and, consequently, were more motivated to pursue the work. It was felt that in this particular topic, which the youngsters themselves viewed as tending towards 'masculine' science and related to 'jobs which men do', traditional intervention techniques directed at the girls were inappropriate.

The physical separation of the girls, in order to give them some form of remedial attention would, it was felt, be damaging to the girls' self-image and would only serve to reinforce any existing preconceptions in the class. Instead, it was decided that an additional activity would need to be devised for the start of the topic, an activity which not only builds up the self-confidence of the girls but which was sufficiently novel that the boys could not complain that 'we know all this' or 'we have done this before'. A second, equally important aim of the activity was that the boys should be aware of the interest shown by the girls, so that they accepted and respected the girls' participation.

The simple introductory activities which were devised were useful in that they guided the less aware youngsters (frequently the girls in this case) towards an appreciation of the use of motors in their everyday lives, while also requiring the whole class to structure their observations in a more formal, scientific way. Each activity can be carried out in under 10 minutes. The youngsters were presented with a collection of illustrations of familiar electrical appliances, all of which used electric motors in some way.

By thinking about the application presented to them, by looking at their similarities as well as their differences, all the youngsters found it

straightforward to identify aspects of the science and technology associated with their design. The opportunity to work from the familiar
(normally domestic) examples was reassuring for the girls in the early
stages of the activities. Later, as their awareness of the scientific principles
behind the design of particular appliances built up, many girls were able
to put aside their earlier views of particular examples as noisy/heavy/dirty
and, consequently, 'unfeminine' machines.

The only materials needed for the exercises are a collection of illustrations of common domestic appliances which contain electric motors.
These examples should be taken from appliances used in the house,
garden and workshop with perhaps a few industrial equivalents (as used
around school). Mail order catalogues are the best source of such illustrations but catalogues for companies which hire tools and advertisements
in the press are additional sources. (Access to a copier which can reduce or
enlarge the illustrations is an advantage.) Individual illustrations should
be mounted on card, as the example below indicates.

Preparation of cards

A sheet of A4 is used to make the 'master' for duplicating the rest of the
cards. The following seven steps are followed:

1. Use a 'borders and layouts' resource to divide the A4 sheet into eight
   sections.
2. Select illustrations and enlarge/reduce to fit area inside border. (If
   illustration doesn't photocopy well it may be necessary to trace a
   line-drawing from the original.)

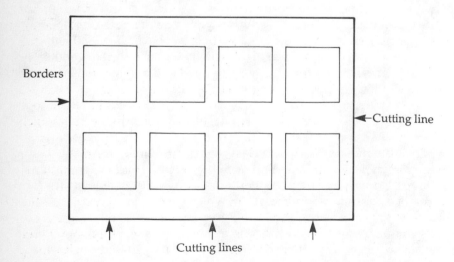

3. Arrange illustrations within borders to make a master copy. Photo-copy and if any of the edges show as lines, blank out with 'liquid paper' or equivalent and use this 'touched-up' version as the master. (It may be necessary to touch up the borders with a thick black felt-tip if the second copy is being used.)
4. Repeat for other illustrations.
5. Duplicate the required number of sheets for sets of cards.
6. Paste A4 duplicated sheets onto backing sheets/card (water-based paste brushed onto backing sheet gives good adhesion when cuts have been made).
7. Cut sheets into eight cards.

To add to the novelty of the situation, 52 cards can be compiled for each group, so allowing the introduction: 'Today we are going to start by having a game of cards.' In this example, eight identical sets were made up. Each set was stuck to a different coloured backing card to allow for easy identification in storage.

Each 'exercise' takes only a few minutes for the youngsters to complete. They are presented as games, and it is possible to fit two or even three of the activities into a 35-minute lesson.

Each exercise requires the youngsters to consider the function of the individual appliances and the basic components from which they are constructed. Then, for example, the appliances can be sorted into groups or families according to their design and/or purpose. The exercises were introduced by means of an overhead projection sheet with instructions written on, but groups could be given sets of workcards containing instructions.

Once their initial reservations had been broken down by handling the cards and talking about the appliances illustrated on them, girls are normally as competent as boys at these sorting tasks and equally imagin-ative in their descriptions of families. Although the tasks suggested to the youngsters were prescriptive, they provided a lot of opportunities for discussion and negotiation of more open-ended work. By the end of the period, some enterprising groups had also devised their own games using the cards and based on variations of 'snap' and 'happy families'. This proved to be good extension work, since establishing rules for a game depends upon having an understanding of the scientific principles of the individual appliances! (This was subsequently attempted as a homework task, with moderate success, but it would be more appropriate as project work.) In addition, there were some opportunities with in-terested classes to develop the activities to consider and discuss the aesthetic appeal of similar items (related to discussion about marketing).

As already stated, the youngsters' attempts at sorting the appliances into families were accompanied by much discussion. Even the youngsters who had not previously shown any interest (in class at least) in the design of any mechanical objects had sufficient experience of the use of the appliances that they were able to make intelligent deductions and were willing to contribute to the group work.

The following exercises have been used successfully with a variety of lower-school classes of different ability as well as with fourth year lower-ability groups following a modular science course. At the simplest level, the cards can be sorted (by individuals or groups) into:

1. 'Ones I have used' and 'Ones I have not used' – useful for the teacher as a way of quickly establishing the degree/range of the experiences of the members of the class; or
2. 'Ones I can name' and 'Ones I can't name' – they should be able to name correctly the large majority of the appliances, but be prepared for some surprises!

Understanding of the design of specific appliances in terms of science and technology can be developed by the following exercises:

1. Grouping appliances which 'work in similar ways', e.g. hair dryers, hot air paint strippers, fan heaters, etc., or
2. Grouping appliances which 'do the same job', e.g. 'cutting' – lawn mowers or strimmers, electric razors, hedge trimmers, carving knives, etc.

If the youngsters are going to devise their own categories for sorting the appliances, then one 'pack' of 52 cards for a group of four is sufficient. Their initial choices of categories are likely to produce lots of small subsets, so 10–13 cards each is as much as they are likely to be able to deal with at first. An integral part of the exercise should be the opportunity to look at and discuss each other's groupings for the appliances. The task can then be developed by allowing the youngsters to work in pairs, combining their cards and trying to reduce the number of subsets needed to group the appliances in families.

Once the youngsters have developed a familiarity with the appliances, their designs and their functions, the following exercises can be used to help them arrange their knowledge and ideas in a 'scientific' way:

- Variations on 'Give us a Clue' where one person selects a card and a second person or a team has to question him/her in order to work out which appliance is illustrated on the card. Suggestions about the format of the questions, i.e. whether they require yes/no answers or are open questions, and any scoring system, may need to be discussed in advance depending on the ability of the youngsters involved to negotiate this at the individual level.

The appreciation of the similarities and differences in the appliances can also be developed by activities to construct simple keys. The advantage of using items represented on the cards is that the youngsters can physically manipulate the cards while thinking through their ideas. The complexity of the task can be controlled by selection of the number of cards involved (say 6 for a first attempt, building up to 10 or 12 perhaps) and/or by specifying particular appliances. If provided with a large sheet of paper, the youngsters can draw the outline of their key onto the paper and attach the illustrations to the backing paper with 'blu-tack', so producing a poster for display afterwards.

The youngsters tackled the activities outlined above with enthusiasm, enjoying particularly the opportunity to think through what they already knew about the topic before proceeding with new work.

The effectiveness of the technique in achieving its aims and objectives is best measured by the increase in confidence shown by the girls in approaching the remainder of the course work on electric motors. In this respect, the technique was a modest success: it was effective for some of the girls for this small area of the curriculum. There was no indication that this confidence was transferable to other curriculum areas, but perhaps this is more of an indication of the depth of the underlying problem rather than a condemnation of the technique itself.

CASE STUDY 20: THE CHERRY PIE GAME

Margaret Davies

Aims and objectives

The aims of games and simulations are to take science outside a laboratory situation; to take a broader view, as experiments in the laboratory tend to be more introspective as far as the student is concerned; to show relevant examples in everyday life or related to world situations; and to place a student in a situation and make him/her think. The immediate objective is to stimulate discussion, because every game and simulation should have time available for a discussion afterwards.

Choice of technique

The technique is chosen to broaden the experience of the student. Science teaching should not be just experiments and worksheets. Games and simulations achieve a greater stimulation of students, as they provide a variation in the type of lesson. They certainly provide an excellent basis for discussion work and more people tend to contribute than they do when a teacher introduces discussion.

Learning group

There are a number of games and simulations already devised for different age groups, but it is often possible to choose one originally written for one particular age group and either extend it for an older group or simplify it for a younger group, while still keeping the content of the game the same. The extension or simplification tends to be mainly in the teacher's expectation. I have used the same game, the 'Global Cake Game', devised for a third year group, for both third year and sixth year students. For this game, a class is divided into groups of different sizes to represent the populations of different countries. Each group has different resources, e.g. paper, scissors or pencils, etc. The groups have to trade with each of the other groups for the resources to complete their part of a wall chart which represents their country's contribution to the cake. At the end of the game a real cake should be produced, cut into a number of slices equal to the number of the pupils in the class. The groups are then given a number of slices, not equal to the number in the group, but representative of that country's share of the world's food/products. Thus one group of eight pupils representing a poor country could get two slices, and another group of two pupils representing a richer country could get eight slices.

The approach has to be varied with different groups even in the same age range. The game or simulation can still be turned to good account even if some of the groups are uncooperative, as the very lack of cooperation can be a focal point of the discussion, e.g. when I was playing the 'Global Cake Game' with a sixth form, a group representing Russia refused to trade and just said *Niet*. This proved an important point of discussion, as refusal to trade brought the game to a halt as it would bring World Trade to a halt. Discussion of dependency on the cooperation of others was the obvious outcome.

Another reason that the approach has to be slightly different is that younger groups do not always listen properly to the instructions, as they just want to get ahead and play the game. They are not bothered about why they are doing it. It is essential that they do listen, so the giving of instructions is more effective in a disciplined situation before the game commences.

Older pupils want to know what is going to be achieved before they start. It is not always possible to tell them this as they will not then fall into the obvious traps and, therefore, the game is spoilt. This is particularly true of some of the survival games which depend on team work, but if the class is informed of this fact, obviously the game will not work.

Pre-planning and organization

The pre-planning must be meticulous. Most of the games and simulations have a time limit and well-listed resources. The resources must be exact and ready. The size and quantity of paper or colour of pencils, etc., must be accurate. The teacher must know exactly what to do so that instructions at the beginning are clear and succinct. It is not the sort of lesson a supply teacher can fill in for, or any teacher at the last minute. Most of all, the teacher must work up an enthusiasm for the experience, because it is an experience for both teacher and pupils.

When one is organizing the class for this type of work, knowledge of the pupils in the class is essential. If a teacher knows that they are capable of dividing themselves into the required small groups, then the class can be just instructed to do this. If the class contains awkward pupils who would stay in one group and destroy the atmosphere, then the teacher must make the division of the class. It is essential to get the utmost cooperation and good will of the pupils for this type of work and, therefore, an imbalance of groups should not be the primary deciding factor.

It may appear that the teacher has a very passive role in between the initial instructions and the discussion at the end of the game or simulation. In fact it is very active. The teacher must observe what is happening in the groups and should be making notes so that, as far as possible, issues can be raised in the discussion. Differences will be noted between different classes and ages playing the same game, so different issues need to be raised. The teacher can also have an even more active role by assuming a character and becoming part of the simulation to provoke different situations. It is possible, of course, to formulate ideas for interventions beforehand. If the simulation flags at all, it is a good extra stimulus. The teacher also has an obvious role of helping any group that has difficulties, although they are usually only asked to perform easy tasks well within their capabilities. It is usually the inability to communicate that would necessitate help.

Follow-up activities

The most important follow-up activity is, of course, the discussion. It is important that the teacher has made initial preparation of the sort of issues that should be discussed, but it is also useful to add any other issues that have evolved during the playing of the game. The primary points could be listed and circulated to smaller groups within the class after playing the game, before everyone gathers together for the final class discussion. More pupils will contribute if they discuss first in small groups. The final discussion will then have more views represented even if there is only one spokesperson for each group.

This system works if the class normally does not contribute much to discussion, or tends to be dominated by one or two people. The other method is to bring the class back into the round at the beginning of the discussion and have a free-for-all.

Other follow-up activities can, of course, be worksheets of questions or fact-finding exercises. In some cases it is useful to use a worksheet before playing the game or simulation as well, to assess how much the students have learned or changed their ideas.

Physical situation

It is important that the students recognize that the situation is not just an ordinary situation but different from the usual lesson. The rearrangement of furniture can usually achieve this but a laboratory is not a good venue for this type of work: the furniture is not always easy to move, chairs are better than stools, and there are far too many distractions, e.g. gas and water taps. A laboratory is obviously not a good venue if any of the games or simulations involve consuming food.

Success of the technique

It is easy to attain the main objective of promoting discussion in class, and it is easy to observe this and the involvement of the students. The games and simulations certainly make students more aware of world issues such as famine, pollution, etc. They also feel much more involved than if they are just given facts. They usually enjoy the experience, but as they get older they do become more blasé. The teacher's enthusiasm is important then to combat any original reluctance to start. Once started they usually continue happily. It is much more difficult to assess how long-lasting the influence of playing the game is.

It would be wise for science teachers to remember that there is an increasing use of games and simulations in some areas of the curriculum. The game or simulation that would appear to fit into a science curriculum could also have an equally valid use in a different curriculum area. Check to see if the game is used elsewhere in the school, because no game can be played twice, however good the motive. The other thing to remember is that whereas it is a valuable experience, the students respond better if they play the occasional game or simulation and are not fed a steady diet of this type of lesson.

CASE STUDY 21: SEXUALLY-TRANSMITTED DISEASES – AIDS

Norma White

This topic, like all other health-related ones, has direct relevance to our lives and requires time not only to understand ideas but also to begin to develop decision-making techniques. It is important to stimulate and teach young people how to make positive healthy choices.

Teaching which involves both simulations and games involves all pupils in exchanging views and ideas with each other. Because this is a much more pupil-centred activity it tends to be less threatening for pupils and this is particularly important for areas such as sex education. In this way pupils not only become more confident with each other but at the same time increase their knowledge, understanding and decision-making skills.

When attempting work on sexually transmitted diseases, it is important to choose pupils who are mature enough to consider the implications of their ideas and new understanding and so begin to translate this into real-life situations. The age of the pupils will vary from school to school but it can be successful with either mixed-ability or mixed-sex groups aged 14 plus. Since there is no individual counselling in the lesson as such (this can always follow on a one-to-one basis later if necessary) the group size can vary from 15 to 25.

There has been much information about AIDS from various sources. What is important is that the teacher's background knowledge is as up-to-date as it can be and there are many resources such as videos available for this. However, facts by themselves are not enough and teachers should, if possible, share their thoughts with other colleagues on such topics as homosexuality, bisexuality, multi–single partner relationships, drug abuse and harm reduction techniques. Statements such as 'You can always tell a gay', 'Enjoy life while you can' and 'How have cats developed AIDS?' encourage discussion. All of this is important in the pre-planning stage. Pre-planning for pupils is important too. They will need some factual information on the male and female reproductive systems, before lessons on AIDS can make sense.

The session itself is really in two halves and this may be of two separate 35-minute periods or one double lesson. In either case the first section should deal with factual information about AIDS and for this I prefer some visual material, ideally slides. So a room that will black out but also accommodate pupils to work in groups of three to six for the following game is ideal.

Young people need time to think about questions relating to AIDS and time to share ideas with peers in a non-threatening way. A card game which promotes discussion and involves making decisions is a useful way to develop this concept. Many readers will be aware of TACADE's (1983), the 'Alcohol – What do you know?' card game, and the one I describe is based on that principle. You need to establish a number of questions relating to AIDS and provide answers to them, and I suggest you make two sets of 10 or 12 questions in each pack. Each set consists of a set of questions numbered 1–10 in the top left-hand corner and another set of answers numbered likewise. A question may read:

- 'You can get AIDS by kissing';
- 'Only homosexuals get AIDS';
- 'You are safer with fewer partners';
- 'Condoms used with a spermicidal cream or lubricant make sex safer from both a disease and pregnancy point of view'.

At least 20 questions/statements are required.

Let us take the first card mentioned – 'You can get AIDS by kissing' – as an example. Underneath this statement are printed the words 'TRUE' and 'FALSE', and beside each of these will appear a number, e.g. TRUE 2, FALSE 8. The pupil who has been dealt this card and whose turn it is reads out the statement and then, unaided, has to decide whether it is true or false. If the answer is thought to be false, the pupil places the card on the table, says 'false' and calls out number 8. The pupil who then has the card with the number 8 in the top left-hand corner reads out his/her card, decides whether the answer is true or false, places the card on the table and calls out the relevant number. If all pupils answer correctly, then a circle of cards will have been placed on the table all following on from each other, *but* if a wrong answer has been given then the sequence will have been broken. At this point pupils look at the cards they were not sure about and discuss them again. For instance, the player may have said that only homosexuals get AIDS. Following discussion the pupils can then refer to the information pack and change the answer if necessary – as in this case. By this process of discussion and clarification the group finally achieves the correct sequence. They can then move on to the second set of questions. Any number from three to six players in each group is appropriate.

The card game is possible in any room setting where pupils can gather in small groups. Pupils in the main enjoy relating the factual part of learning to situations of everyday life. These activities help the young to talk about such issues which most of us, to a greater or lesser extent, find difficult. In this instance they are also learning to evaluate the relationships between some very powerful life-events – risk, pleasure, sadness, death. The hardest part for the teacher is to promote the idea of giving the young time and space to think, without being threatening.

AIDS is a threatening disease for two main reasons:

- at this stage it is incurable; and
- it requires a possible realignment in contraceptive and general sexual behaviour.

This card game provides some kind of opportunity for young people to discuss facts, myths and situations. Follow-up activities can either be in the next lesson or in some other curriculum area, such as home economics, physical education or religious education.

Some suggestions for follow-up science topics might include bacteriology, immunology/vaccination. Other important learning approaches when dealing with this topic include, for example, the media coverage of AIDS and role play, perhaps based on fears and anxieties of being a blood donor, going to parties, being heterosexual or bisexual and looking at the relevance of human seduction in sexual behaviour.

How successful are such techniques? Only the pupils' evaluation of your approach can answer how successful you are. Because I find sexual issues so important for the pupils' social development I have always asked the question, 'Has this work helped you to think and consider, or has it put you off for life?' My information to date is that this kind of approach helps the young to think, consider and begin to mature.

CASE STUDY 22: COMPUTER-ASSISTED LEARNING – A BIOLOGICAL SIMULATION

Anita Pride

Deciding to use a computer simulation as a basis for a series of lessons in a fourth year biology course seemed, at first, to have solved two major problems. It helped to find a way of extending ideas about how organisms interact with each other and with a complex and varying environment, while at the same time the setting for these interactions showed something of the economic and social pressures they could exert on people (populations?) in developing countries.

This particular computer program, which closely models the incidence of malarial infection in a remote African village, presents the children with a very real problem. They were cast in the role of medical control officer, who had to reduce and then control the number of villagers infected with malaria by mosquitoes. They had to effect this control as cheaply as possible using the limited resources available to the village people. The effect of such a technique was to bring a physically remote and unfamiliar situation close to the children, making them personally involved, as they made decisions about which of the six control measures to try, implemented them and saw the results of their actions.

The interactive nature of the approach was a principal reason in its choice. Understanding how hard it can be to control any situation that includes a wide variety of variables, such as people, insect populations, infectious organisms, changing weather, education, housing and money, is a tall order for anyone. Being able to choose and make changes in this complex situation and see immediately the response provides a way to help youngsters towards that understanding.

Another important feature of the approach is that it provides success for individuals at many different levels which is a very desirable quality when working with a group with a wide ability range. Some of the children chose impulsively and worked randomly; they were able to see that infection levels had fallen (or risen) under their management but could not pin down a reason for it. Others were able, after a few tries, to see the value of planning their strategy carefully. They were able to use their strategy to answer their own questions: questions such as, 'Is there one control measure which will work on its own?' The cautious ones searched patiently through the database of records from other villages, looking for a basis on which to make their decisions.

The first session was organized in the school library, where pupils saw a video film on relationships between organisms. Part of this showed scenes from an African village where malaria was endemic. This provided an important stimulus in helping pupils to visualize the village environment and see how control measures were carried out in actuality. It helped pupils to contextualize their learning in reality. However, although extremely valuable, the film is not essential, since the computer-assisted learning material can stand alone. Pupils did need other background information, however, such as photographs of *Plasmodium* cells, mosquito specimens and information about life-cycles and climate. These provided important ways of helping to answer questions and resources for pupils to seek their own answers.

After this introductory half-hour, the group moved to the laboratory where the program was quickly and easily loaded into the microcomputer as the children gathered round the largest monitor screen available. The first 10 minutes consisted of teaching everyone the main points about how to run the program and which keys to use. Pupils' questions, such as how to move from one part of the program to another were answered at this point, although they obviously came up later as well. Their job as medical control officer for the village was presented to them and they were given copies, taken from the students' booklet, of the village map, along with some basic information about how mosquitoes spread malaria. The session ended with a general talk about the ideas they wanted to try out next session when their turn at the microcomputer came round.

The next session began with the sorting out of small working groups of two or three. Each group then planned their strategies. This consisted of both listing a sequence of actions they could take to answer their own questions and devising a chart or table on which to record their actions and what resulted from them. Since only one machine was available, each group was allowed about 6–7 minutes to put its plan into action; this gave them enough material for useful work away from the microcomputer. (If an econet version and a computer room had been available, then far more hands-on time would have been possible.) Trying to explain the reasons for what happened, for their successes and failures, were the most valuable learning exercises. One vital message came over to the children quite clearly, as each group's progress report appeared on the screen at the end of a 6-year term of office: being in charge means being responsible for what happens.

One feature of the technique became very evident: the small groups, as they came in turn to the microcomputer, tried out arguments, bounced ideas off each other, put forward plans and had to back them up and, in general, almost all took a far more active and purposeful part in the lesson than they usually did.

Two sessions, each of 70 minutes, were spent in this small group work, giving each one a chance to try two full terms of office. Homework tasks during this period included organizing the results to present clearly to the class (some as overhead projector transparencies), and writing a concise summary of what they had achieved, with reasons suggested for their success or failure to control the disease. Some children prepared plans for yet another try!

A final session, taken in a classroom with no computer, was used to prepare sections of a script which was later used to make a 'radio programme' on audio-tape. Their experiences with the computer simulation, the interrogation of the databases of other village records, and their reading and discussion of the printed material were all used eventually to produce a tape based on the pupils' taking roles in the village. They then produced a tape of four short episodes, based on the stories of four people: a child or mother in the village telling what life is like there and the problem they having in avoiding mosquito bites; a previous (retired?) medical control officer talking about his term of office; a health worker explaining why we, in England, do not have these problems; and a WHO officer telling of places in the world where malaria has been beaten and how it was done.

To sum up, this approach to creating a learning situation in science has a number of facets which should persuade any teacher to attempt its use, in spite of the many practical difficulties involved:

1. It stimulates interest in the area of study by enabling the child to take on responsibility for his or her own learning.
2. It opens up opportunities for language use and communication with a purpose.
3. It facilitates the reviewing, sifting and recognition of relevant data which can contribute to decision making.
4. Specialist knowledge and organized data can be drawn upon freely and in a fruitful way.
5. There are chances for designing an investigation, finding faults with the design and then making changes, with no waste of materials and a minimum waste of time.
6. The power of the computer is used to make redundant any tedious and difficult calculation or eliminate long periods of waiting, so that the overall concepts become much more accessible. There are opportunities for the children to select their own individual pathways through the same material and yet make progress up the same developmental curve.
7. Since the material for more computer-assisted learning programs has already been well-researched and presented, it frees the teacher from a purely didactic role, to perform in a more subtle way. There is time to recognize frustration, to stand back and listen so that you know when to help and when to keep quiet, to encourage self-direction but offer guidance and to learn along with the children.

SUMMARY

The four questions we normally address at the end of each chapter concern active learning, prior preparation, organization and other areas of importance. We feel that many of the aspects of organization have been covered more than adequately by our contributors, and thus propose to consider only three in this summary – active learning, prior planning and other important aspects.

1. Which aspects of active learning do they encourage?

Making decisions, solving problems and displaying understanding and competence in different ways are all assisted by games and simulations.

So too are autonomy, personal investment, and motivation. Perhaps because games are such a feature of learning experiences outside school, many pupils enjoy them enormously and find learning much less effort when it is couched in this form. Pupils usually respond well and enter into the spirit, without overplaying the competitive aspects of gaming.

2. What planning do they require?

Many games can be taken from published materials. Others, based around common gaming formats such as snakes and ladders, snap, dominoes, happy families, etc., can be adapted by teachers to match their needs, as case studies 19 and 21 clearly show. Simulations of the computer type are also common, though unless you are very skilled they are probably not easy to provide yourself. However, there are many more simulations than just those on a computer. Ken Jones (1985) has some excellent advice on designing these kind of simulations. He calls his approach to design:

> the cooking pot approach. The analogy is that of a chef who stirs the pot and tastes the mixture, and adds or reduces ingredients. It is the blend and nourishment that matters. The criteria are plausibility rather than reality, consistency rather than sequence and interest and challenge rather than modelling.

He outlines four basic questions essential to the process of designing simulations. It is these key questions that we feel are important steps for anyone undertaking gaming and simulation in their classroom. It is too easy to regard games solely as interesting and entertaining activities and make their role in lessons one of 'breaking the monotony' when, in fact, if carefully organized and skilfully planned, they have much more potential to assist the learning of students. Jones' questions are:

- What is the problem?
- Who are the participants?
- What have they to do?
- What do they have to do it with?

All our case studies have addressed, if not overtly at least implicitly, these questions. They have outlined specific educational problems which require a solution, chosen a problem-solving example with which students can engage, defined their participants, organized the rules, and prepared the resources.

The questions are important ones for teachers who are considering designing their own games and simulations. They represent, for us, some of the fundamental basis for planning.

3. What else is important?

One of the most important aspects of simulations has, so far, been left relatively unaddressed. That is the role of computers in the simulation of experiments. Computer simulations can enable the understanding of experimental processes when 'the experiment is too complex, expensive or dangerous for student use; experiments where the apparatus is notoriously unreliable or where the experiment would take too long' (Masterton and Chaundy, 1978). Moore and Thomas (1983) see computer simulations of experimentation as having a great deal of value. They claim that 'because experimental variables . . . can be changed quickly and often, many pupil suggestions can be taken up . . . and investigated in a short time'. In other words, computer simulations of experiments can aid active learning by giving students more control over the conditions, and removing the necessity of spending large amounts of time setting up apparatus.

Jenkins (1987) claims a more fundamental importance for the role of computer simulations. He states that they enable students to develop scientific models, and use them to test out hypotheses. Given that one of the primary functions of science is to produce explanatory models of how the world works, then computer simulations obviously have a very powerful role to play in helping students to acquire the skills of modelling.

Several authors, including Jenkins, warn of the danger of replacing experimentation with computer simulation. He claims that to develop models and then use these to test out hypotheses without experimentation is not scientific. Further, he states that the model will be a simplified version of reality. We would argue that all models used in science to explain the world are of necessity a simplified version of reality. Choosing to control one or more variables and ignore others out of a wide variety of possibilities is equally simplistic. Having access to equipment does not necessarily argue for the complexity of science or scientific process. It may well be that pupils can develop skills of hypothesizing, testing and revising their models to a fairly sophisticated degree by the use of computer simulations in a way that 'real' experiments make much more difficult. If computer simulations aid the understanding of the basic ideas of science then that has to be, in our opinion, an excellent reason for their inclusion in the curriculum.

BIBLIOGRAPHY

Assessment of Performance Unit (1985). *Girls and Physics*. London: DES.
Brandes, D. and Phillips, H. (1979). *Gamesters' Handbook*. London: Hutchinson.

Jenkins, D. A. (1987). The use of computers in the teaching of biology. *School Science Review* **68** (245), 687–93.

Jones, K. (1985). *Designing Your Own Simulations*. London: Methuen.

Masterton, D. and Chaundy, D. C. F. (Eds) (1978). *Computers in the Curriculum Project: Physics*. London: Edward Arnold.

Moore, J. L. and Thomas, F. H. (1983). Computer simulation of experiments: a valuable alternative to traditional laboratory work for secondary school science teaching. *School Science Review* **64** (229), 641–55.

Rice, W. (1981). Discussion, role play and simulation in the classroom. In *Health Education in Schools* (Eds J. Cowley, K. David and T. Williams). London: Harper and Row.

Teachers' Advisory Council on Alcohol and Drugs Education (TACADE) (1983). *Alcohol – What do you know?* (Tour of Knowledge Card Game). Manchester: TACADE.

7: USING ROLE PLAY AND DRAMA IN SCIENCE

INTRODUCTION

There is a hefty literature on role play and drama in schools, though, significantly, very little on role play in school science. On the whole it is not something which is familiar to science teachers. For this reason it might be worth spending some time in exploring the area. Many science teachers might be sceptical of role play and drama in science. How, they might ask, can factual, conceptual, practical science be made into drama? What is the point of a youngster trying to role play situations in physics or chemistry?

Those with some experience of the real virtues of drama and role play would approach the issue from another direction. Over the years there has been a small but growing band of science teachers who have come to appreciate its attractions. Many will remember that The Molecule Club at the Mermaid Theatre in London's West End has a long and honourable history of enacting scientific ideas, and in involving youngsters in dramatized science. Recently, through the University of Wales at Cardiff, there has been a national science drama competition (Eisteddfod) drawing contributions from schools from all stages and places. The Annual Meeting of the Association of Science Education at Cardiff in January 1987 staged an excellent symposium on drama in school science; Charles Taylor (1987) develops the trend in *Physics Education*.

Though this may represent what is tantamount to the grand, public face of science drama, at least it signals that it is beginning to be taken seriously. Much of what actually happens in classrooms is much more low key and informal. Early attempts to bring some active pupil participation to science lessons were the kind of activities first introduced in the Nuffield schemes. In Nuffield physics, for example, when introducing the ideas behind kinetic theory and molecular motion, youngsters are asked to stand in line – arms linked – to simulate the motion of particles in a solid. Enacting molecular motion was a lively way of helping understanding. Our case studies in this chapter do not explore physical enactment of scientific processes by youngsters, some of which were tackled in the last chapter. Playing at being a molecule, though, has been one of the more common methods of active modelling in science. Here we concentrate on educational dramas, and role playing within them.

It is important to note that role play and drama are significantly different activities. We do not want to spend time forging esoteric distinctions, but it might help to separate the two. Essentially, it is a difference of personal investment. In educational drama, youngsters are asked to explore situations. They may be asked to be research scientists, historians or manufacturers investigating the feasibility of a scientific idea. They may decide to be gruff and authoritarian if, for example, that is the way they see scientists or researchers to be. But the role itself is not the main point of the exercise. People are usually asked simply to be themselves in a different situation.

In role play there is much more empathy for a kind of person, or particular context. An example might be the work undertaken in history lessons where the skills of empathy are encouraged in specific historical contexts. The role becomes important as a way of exploring ways of working, thinking, feeling and behaving. In some cases people will play a specific role within a piece of educational drama – this is the case in each of our three case studies here.

The three contributions are similar and yet very different. All three dramatize situations and then ask youngsters to carry through with the 'scenario' to explore the various routes and avenues open to them. Jon Nixon describes a lesson that introduces youngsters to biotechnology. The class are asked to take on roles in marketing teams brought together to design and sell a new brand of yoghurt. The teacher, and perhaps a colleague, have specific roles to play as the marketing director and a health specialist. The roles taken by the pupils are much more open-ended. Martin Hollins writes about challenging youngsters' understandings of mechanics. In one lesson, he and a colleague operate as Galileo's inquisitors, while the class are asked to be twentieth-century defenders of his theories. In another, Martin is an energy minister and the class are scientific advisers, all concerned with energy conservation. Hamish Fyfe's description of a genetics lesson involves youngsters in a drama in which he and a colleague play particular roles. The youngsters are asked to be television researchers preparing a programme about Down's Syndrome. The lesson invokes a range of moral and ethical issues as well as prime scientific concepts. As such, the substance of the drama has the possibility of strong emotional overtones. In some respects the point of the drama is to explore the situation; in other respects it is to explore the human and social implications. Although the youngsters are not 'in role', there comes a point in the drama when as a result of one boy's question, they empathize strongly with the mother and child and this leads to a greater involvement emotionally for the youngsters.

What are the reasons for choosing to explore science through drama and role play? Our three contributors lay out their reasons very clearly and we are hard pushed to improve on these.

First, and most important for us, is that it provides an excellent way for the teacher to both retain control of the lesson and its focus, while at the same time devolving responsibility for learning to the students. It allows the learner, as Hamish Fyfe says, to come into active relationship with the material.

Second, drama and role play allow youngsters to make appropriate judgements about attitudes, values and feelings, as well as facts and concepts. As noted before, this has been a neglected area of science education and these kinds of techniques are good ways of providing a context for discussion.

Third, as we have noted for other strategies, role play and drama encourages oral communication. For example, it is a way of enabling students to put half-formed ideas into words, to try out fledgling ideas, to test their 'thinking-not-yet-finished', as Jon Nixon calls it. Moreover, it is a way to help them develop arguments from other points of view, and a way of presenting their work to a wider audience.

Finally, it builds on children's experience, and allows them to relate their experiences to the outside world. It is motivating, as in Martin Hollins' case, even for reviewing and revising work. Importantly, it is a way to explore the social and personal implications of the science they have learnt.

CASE STUDY 23: DRAMA, SCIENCE AND ISSUES OF VALUE

Jon Nixon

Arguments for the use of drama as a mode of learning across the curriculum have been well-rehearsed over the last 10 years (see e.g. Nixon, 1982). In practice, however, the cross-curricular function of drama has, to a large extent, been limited in its impact on English studies (e.g. Evans, 1984) and the humanities (e.g. Dodgson, 1984). While it has made a significant contribution within these curriculum areas, its potential as a teaching tool within other curriculum domains has been somewhat limited.

There are, it should be acknowledged at the outset, limitations on the uses of drama across the curriculum and particularly, it might be argued, within the science area. No-one would claim that drama should be the only teaching method employed. However, as one among a range of strategies, it can be highly effective in a number of ways. It can, for example:

1. Raise students' awareness of the value issues associated with certain topics.
2. Allow students an opportunity to develop arguments from a specific perspective.
3. Provide them with a means of presenting their work to a wider audience.

Central to each of these functions is the notion of 'role'. Put very simply this means that, through drama, students are able to explore ideas, opinions, viewpoints and attitudes by *pretending* to be either someone else or themselves in a very different situation. That is, drama can be a way of trying ideas out, of looking at problems from different perspectives, and working collaboratively towards a provisional solution: a kind of 'thinking-not-yet-finished'.

How does this work in practice? In order to explore the practical application of drama within the science curriculum, let's take a look at a mixed-ability group of third year pupils involved in an integrated science course within an inner-city mixed comprehensive school. Within the course is a recently devised biotechnology component of about half a term's duration.

One of the units within this component involves groups of students in sampling a variety of commercially produced yoghurts and comparing them critically with regard to a number of factors, e.g. colour, consistency, taste. Drawing on the findings of this exercise, each group of five students then devises and produces its own 'ideal' yoghurt by selecting from a range of ingredients supplied by the teacher.

It is at this point – having prepared the ground in previous lessons through small group discussion and empirical enquiry – that the teacher shifts the learning into the dramatic mode. The lesson begins with each group being asked to imagine that it is a marketing team brought together with the express purpose of selling the newly devised yoghurt on the commercial market. They are to pay special attention to:

1. The kinds of arguments that might be used in support of the new brand (e.g. health, price, taste).
2. Specific strategies that might be used to support its sale.
3. A means of evaluating the product in terms of the arguments developed in (1).

Each group is also asked to elect a general spokesperson and to designate specific areas of concern (e.g. health factors, cost-effectiveness, packaging, marketing strategies) to particular members. A 'meeting' (chaired by the teacher in his/her role as the marketing manager) is then convened to discuss the merits of the various brands and to select one of these for intensive promotion. Each of the groups participates in this relatively formalized role-play, with general spokespersons offering the broad arguments and other members supplying more detailed information and supporting evidence.

Clearly, such a scenario is, on the face of it, mechanistic, functionalist and supportive of a particular value orientation. The role the teacher adopts is, therefore, of vital significance in raising alternative issues of value, teasing out divergencies of opinion within the group, and pointing to different orders of evidence. Establishing such a role depends upon posing carefully prepared questions and defining beforehand a clearly articulated issues-focus. (Other more practical points also require some prior consideration, of course; for example, the use of space in the specialist science area and the safety factors involved in movement around the laboratory.) The teacher-as-chairperson is in an ideal position to question, prove and challenge the assumptions of the group. This requires detailed planning: though 'spontaneous' in form, drama is perforce – in structure – highly patterned and premeditated.

In this particular lesson the role of teacher as prompt and challenger was supported by the presence of a co-teacher who adopted the stance of a somewhat vaguely defined 'health specialist'. Throughout the 'meeting' this teacher doggedly referred the group back to value issues relating to health and to the question of social responsibility within the commercial sector. In a different context a student might well be selected to perform this role, provided, that is, that a student with the necessary social and political awareness could be located within the group. In the absence of such a student, or co-teacher, the teacher could still challenge the assumptions of the group through careful questioning and the presentation of alternative viewpoints.

The final phase of the lesson was again couched in the dramatic mode, but students were this time asked to work in pairs. The drama was set several days after the meeting and involved each pair in acting out a conversation between a member of the marketing team who attended the meeting and a close friend who was not present on that occasion. Those taking the role of friend were briefed separately by the teacher during the lesson and asked to find a way of challenging the position adopted by their partners. This strategy aimed at providing a further opportunity for critical reflection on the issues.

Clearly, time is a crucial factor in work of this kind. The teacher in question was fortunate in having sessions of 2 hours' duration, which allowed both for in-depth enquiry and for the use of a wide range of tasks and strategies. The drama was thus placed in the context of the students' on-going investigative work within science, rather than constituting a discrete component that they might have found difficult to relate to the rest of the syllabus. Nevertheless, it should be noted that drama can still make a significant contribution, even under less favourable circumstances than those pertaining in this situation. Though flexible and generous timetabling facilitates and sustains a cross-curricular approach to drama, it is by no means the only, or even the most important, factor. Informal links between colleagues, a willingness to adopt an enquiring and experimental approach to teaching, an urge towards collaborative teaching, etc., provide the real basis for the development of drama within the science curriculum.

Within this briefly sketched lesson, drama was used almost exclusively within the *process* mode; as a means, that is, of explaining issues implicit in the findings of the students' empirical enquiries. It should be noted, however, that drama also has an important part to play in the presentation of such issues to a wider audience – in the articulation, that is, of *product*. The use of school assemblies, year-group meetings, even parent meetings, can be a useful adjunct to the link between drama as a mode of learning and science as a mode of thought. Drama is a communicative as well as an expressive medium.

A common criticism of drama – sometimes offered as damning praise – is that it provides an opportunity for the 'less able' or 'less academic' student. Drama, however, is a discipline of thought in its own right. Operating at both the social and cognitive levels, it provides students with a space within which to link 'scientific' and 'spontaneous' concepts. It traces the continuities between the more formal aspects of the science curriculum and issues relating to the social, personal and political development of the individual. As such it has a vital role to play across the whole 5–18 school science curriculum. Far from being the preserve of the extrovert, drama is potentially a deeply reflective process to which *all* students can make a unique contribution.

Finally, it should be noted that the kind of cross-curricular perspective advocated in this brief report on drama as a mode of learning relies on strong in-service support for the professional development of teachers. If drama is to be used as a teaching tool within the science curriculum, then science specialists will need to be trained – and drama specialists possibly re-trained – in its cross-curricular application (see Nixon, 1986, 1987). This development challenges practitioners, policy makers and those responsible for both the initial and in-service education of teachers.

CASE STUDY 24: EDUCATIONAL DRAMA

Martin Hollins

Two examples of the use of role play in fourth year science classes are described. The technique of educational drama has considerable potential in teaching science, and here it is used to enable pupils to explore the implications of scientific discovery and technological change, and to review their own understanding of scientific concepts.

Introduction

Role play offers an effective and economical way into the use of drama in the classroom. It builds on pupils' general experience (they have roles as students, as sons or daughters, etc.) and knowledge (in this case, of science). This is then set in an unfamiliar context in which the student is given a particular function, by the teacher also in role. The process of the role play is the student behaving as closely as possible in the way in which he/she feels is appropriate to this function and context. This presentation is for the student and fellow role players, not for an audience. This encourages the use of oral communication skills but does not require the performance of an actor.

Role play can be effective in motivating students, in relating the experience of the classroom or laboratory to the world outside and in providing a safe opportunity to display attitudes and feelings. Role play is economical because, though it demands some careful planning by the teacher, it requires no preparation by the students, no special facilities and the minimum of 'props'.

Planning

Cecily O'Neill and I, respectively Wardens of the ILEA Drama and Tape Centre and North London Science Centre, were invited to use such an approach by science teachers of an inner-London comprehensive. After discussion, we decided to intervene, in role, in two fourth year classes studying integrated science. Each group included pupils with wide-ranging abilities in science, both in articulateness and self-confidence. One group was coming to the end of a topic on mechanics and astronomy and the other, which we met later, had been studying energy resources.

The use of drama was intended to achieve the aims of the science course, not have any specific purposes of its own. In particular, the teachers wanted the students to (i) review and revise work just completed and (ii) explore the implications, both socially and personally, of the science they had just learned. Role play offered a novel way of revising and the freedom to express views on implications. With the class teachers we discussed possible approaches suitable for each group and brainstormed ideas for scenarios. A direct challenge to conventional theories of mechanics seemed likely to provoke a response which would show how much the students had understood. As they had also done some work on Galileo's conflict with church authorities we decided to take up the role of Galileo's inquisitors, but at a later date than his well-known trial, so that there was more freedom for invention. For the energy resources group we decided on a more technological–political approach, setting the scene in the future to give freedom for imagination so as to distinguish the exercise from straightforward problem solving.

With these half-formed plans in mind we went away to do some background reading, especially on Galileo's life. Historical accuracy was not essential but we needed to avoid any distracting argument about the credibility of our role situation. We returned to the teachers for a final planning of each lesson. We considered carefully how the teacher would introduce us, how we would set the scene and how we would develop the role-play during the 50-minute period. We also considered how to end and what follow-up the teacher would carry out.

We also decided that the students would be given no notice of what was expected of them, their teacher saying only that there would be visitors to the lesson in question. This avoided any worries that the pupils might have had about having to perform: in one lesson in which the teacher revealed that there was to be some drama (as the pupils were assembling), one pupil responded, 'Oh no, I've got to work for my exams!' In the circumstances, she had no time to develop this protest, and soon became an active participant.

The lessons

These took place with the students sitting around the benches in their normal laboratory. There was no physical action in the role play, apart from forming groups and presenting findings. In each case we started with the teacher simply introducing us as visitors. We then introduced ourselves in our roles, giving information to set the scene, making clear the role of the students, and starting the action with some carefully provocative questions.

It is appropriate here to describe the two sessions separately.

Galileo's manuscript

An influential contemporary of Galileo, the Duchess of Siena, has discovered that he has broken the injunction laid on him at his trial and published a full account of all his work. This includes his revolutionary ideas in mechanics and his belief that the earth is not the centre of the universe. She has come forward in time to ask the class as twentieth-century scientists, whether this work has any value, and what punishment is appropriate for his devious behaviour. She has with her a 'clerical scientist', Brother Martin.

Our attack was both scientific, with Brother Martin espousing the 'common-sense' Aristotelian views of projectiles, etc., and political, with the Duchess expressing concern at the dangers of freedom of expression leading to anarchy.

The result was very stimulating. The students attempted with ingenuity to demonstrate to us the technicalities of their mechanics experiments. They hotly defended Galileo's, and anyone else's, right to freedom of expression and went on the attack against the repressions of the church at the time. Such was the pace of the exchange that we did not find the opportunity to break into groups to allow the students to produce a more considered defence. As time ran out they realized that they had not convinced us technically, and were beginning to wonder whether today's total freedom was altogether for the good of humanity! The other disadvantage of this continuous exchange was that only a minority of the students spoke, though the attention of all was held. We agreed to try to return, after some preparation on their part, to finalize our assessment of Galileo's work and to announce our verdict on him.

Energy policy 1996

As government energy ministers we informed the class of the serious energy shortage consequent on the total loss of generating capacity by oil or nuclear fuel in 1996. As government scientific advisers, their task was to devise technical solutions *and* the necessary measures to persuade the public to adopt the changes in lifestyle that were essential. In the opening session of about 15 minutes we made some suggestions and drew some others from the class. We then divided up the task into aspects such as housing, transport and industrial uses, and asked for groups to form and choose one aspect to work on in detail.

Groups of between two and five sorted out their own ways of working for the next 20 minutes. A few pupils drifted from one group to another, and one group was a collection of individuals working independently. Most, however, quickly sorted out the issues they wished to address and tackled either technical aspects, such as designing energy self-sufficient dwellings, or politico-economic ones such as the control of private transport, or both.

Each of us, and the teacher, visited the groups, prompting with ideas or information as appropriate, in role. We then asked for reports from each group which could be questioned by members of other groups. Though time had been very limited all groups produced some interesting ideas, and many showed an integration of science into more general views about lifestyles that was quite impressive. The role play ended with us thanking them all for their contributions, while highlighting some of the social issues that they had raised.

Follow up

In both cases homework was set for the students to draw together some of the ideas which they had expressed and been exposed to from our introduction and the class development. In the first case this was to be used for discussion at our return visit, but this proved impossible before the end of term and so the response was turned into a letter to us defending the particular case of Galileo, and the general issue of 'scientific freedom' (an extract of one is shown in Fig. 24.1). In the second case members of the groups which had formed, refined their particular 'solution' to the problem of energy shortage, which led to an interesting mixture of technical design, social policy and advertising copy! These products were discussed briefly by the teacher in a subsequent lesson.

Evaluation

Our assessment of the success of these two brief experiences is based on our impressions at the time and the teachers' observations at the time and immediately following.

High levels of motivation were generated at the time, and this carried over for a short period afterwards, so that the students produced some good written work as follow-up. Students became aware of shortcomings in their understanding when trying to develop an argument to persuade others (both ourselves and other students). They showed considerable commitment to their views, sometimes quite clearly in an adopted role, and drew on a whole range of experience outside of science in formulating their views. For example, in 'Galileo', one student used his knowledge of medieval church practices to accuse us of corruption in the selling of indulgences. In 'Energy Policy' many students considered technical solutions in the light of what they thought would or would not be acceptable to the public, and what was justifiable behaviour in attempting

Dear Duchess and Brother Martin.

Due to the fact that you will be unable to join us to for further descussions, I am writing to you to explain some of the things we might have talked about to try and convince you that the great scientist Galileo's acusations are true. For cinstance his theories about motion without force. In the 20th Century we have quite advanced technology. So we are able to prove his theories.

This is a linear air track. This is used to portrail the conditions without friction.

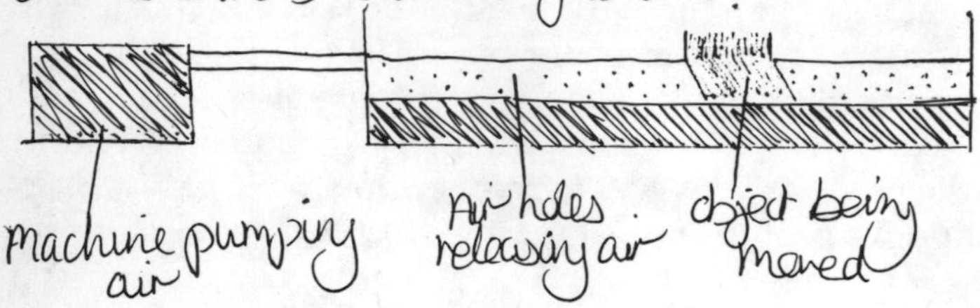

machine pumping air   Air holes releasing air   object being moved

Air is pumped through the pipe to the apparatus and out through small holes which act like a cushion of air which allows the object to be moved along with ease (without friction).

When you push the object it travels at the Same speed without stopping and would carry on doing so if the track was longer.

This proves that if no force is acting on it an object carries on moving at a steady speed in a straight line, or it stays still.

Tracey.

Figure 24.1

to change public opinion. We also felt we had succeeded in encouraging some students to be creative in their responses to both our ghosts from the past and our spectre of the future.

In conclusion, we considered that the use of this dramatic technique had achieved some of the aims of the students' science course which would have been difficult to achieve as effectively and as enjoyably in other ways.

CASE STUDY 25: 'X, Y AND Z' – A DRAMA-BASED APPROACH TO THE INTRODUCTION OF GENETICS IN THE SECONDARY SCHOOL

Hamish Fyfe

Using drama as a strategy in the teaching of science may seem an unlikely ploy, but as Gavin Bolton (1984) points out in his excellent book *Drama as Education*, such approaches are far from new. Bolton quotes Miss Harriett Finlay Johnson writing in the first decade of this century:

> It may not be the facts themselves which are so valuable. It is the habit of mind formed whilst learning them which makes their worth.

This statement would not be out of place in a discussion about the curriculum of secondary schools today and provides an indication of the philosophy which guided the teaching team in its work on genetics which is described here.

Aims and objectives

We aimed to introduce concepts of and ideas about genetics by employing an imaginary 'drama' context. It was hoped that the context would provide the children with a set of parameters within which they would explore issues and discover facts. The pressure for this exploration would come directly from the imagined context. The whole group, including the teaching team, were involved 'in role' in the drama throughout. By 'framing' the learning group as documentary film researchers the children viewed all the feelings and facts presented to them from a particular perspective, i.e. whether the material presented was suitable for inclusion in a 30-minute television programme on the 'human issues' of Down's Syndrome. We hoped that working in this way would allow us to move away from considering the children as 'empty vessels' –

which the teacher would attempt to fill, at least partly, with second-hand information – into an active relationship with the material. Thus, once the 'big lie' of the drama had been accepted, the children were put in a situation in which they had to search out and make appropriate judgements about facts *and* feelings.

The learning group

The group described here were a fourth year group in a Belfast secondary school. They were in the early stages of a CSE biology course and were considered to be the 'less able' of two such groups. It was hoped that some of the difficulties which might lie in obtaining whole-group discussion might be overcome by using the dramatic context to exert a series of pressures on the children's use of language. The drama was designed to provide a focused *context for discussion*. The class's new role as television researchers placed specific demands on their language as they met and interviewed doctors, parents of children with Down's Syndrome and accommodated the demands of the director of the programme, all within the fictitious context of the drama.

Pre-planning

Detailed planning took place between the class teacher, college tutor and B.Ed. students, all of whom were involved in the project. These circumstances are different from those normally experienced in planning for work of this sort but the team felt that a great deal was gained by the fact that the teaching was a cooperative endeavour. It was felt that such a team could be created, admittedly with effort, by involving other members of staff, whether science staff or not, older pupils in the school or even parents. A large team, however, is not essential to the strategies which are described here.

The team felt that the tasks which researchers have in preparing material for documentaries was likely to produce the sorts of pressures on language, deductive and decision-making skills which were appropriate in learning about genetics. The team prepared a letter from an imaginary film company appointing the researchers and inviting them to an initial meeting with the director. With one of the team working in role as the director, that meeting was to provide the initial impetus for the drama.

All the researchers (a group of 17) were provided with folders containing a 'pack' of information about genetics, some from scientific journals, and many press cuttings pertaining to cases in which genetic problems had led to a high 'human interest factor' for the media. The learning group were thus assumed by the fictitious context to have skills as researchers but *not* to have any more than a general knowledge about genetics. Once the initial meeting with the director of the programme was over, the researchers were to have a number of roles available to them to discover whether what they would have to say was appropriate for the programme.

It should be stressed that all members of the team were clear that the teachers' adoption of a role *did not mean acting a part*. Peculiar clothing, odd accents and eccentric behaviour would all be a distraction from the task in hand. The role of the director of the television programme, for example, did not differ greatly from that of the class teacher. He was taking charge of the work of the group, ensuring that it completed tasks, and chaired all discussions. He presented a series of attitudes towards the creation of the programme and the planning team decided that he should place pressure on the research team by constantly trying to sensationalize the issues which were being dealt with. This, as was hoped for in the planning stage, did produce a strong moral reaction from the rest of the group, and focused the learning for a time at least on the distortive power of the media.

Other roles created for the session included the mother of a Down's Syndrome child and a medical 'genetics counsellor'. It was decided that the researchers should meet and interview these people, without the director, to assess their potential contribution to the programme. A specific but limited set of circumstances were created for the mother including name, age, marital status, and so on. Other than that it was agreed that her attitude toward the problem would be a positive one, i.e. she was considering having another child and was not overly concerned about the possibility of its suffering from Down's Syndrome. The doctor, on the other hand, was concerned that 'preventative measures' should be taken in order to avoid bringing into the world children with severe physical and mental handicap. It should be stressed that this was in no sense a script and that the teachers adopting these roles would respond directly through them to the needs of the children as they perceived them at the time. The option exists, as it does of course in real life, not to answer indelicate or inappropriate questions. A real example of this arose during the teaching when a member of the research team asked the teacher in role as the mother how long she expected her child to live. The mother's unspoken response, the silence which followed, the reassurance and the careful rephrasing of the questions deepened the experience for all the participants.

It was decided that the planning stage to present the researchers with the task of explaining as succinctly as possible, and to the general public, the 'mechanics' of the genetic transfer of characteristics, as an initial part of the programme. The class teacher, using a specially prepared visual aid, was introduced by the director as an expert to outline basic principles. The research team took notes and then discussed various methods of relaying the information to the general public. After some detailed discussion with the class teacher, a particular sequence of computer graphics was chosen as being most likely to be clear and of interest to the general public. The design of this sequence was left until a later stage but seemed likely to involve equipment already in the school.

The organization of the session

The session took place in the group's usual science classroom which had permanent benches.

One of the teachers explained to the class that they were going to do a play about genetics, but that they need not worry about acting or being stared at because everyone, including the teachers, were going to be in the play at the same time. The roles which they were going to meet and the fact that they were going to be researchers of an imaginary television programme was also explained. Time and care was taken in starting the drama to ensure that the children had time to negotiate, in context, what being a researcher actually meant and to ensure that belief in their role would build gradually rather than simply being imposed by the teacher. The discussion about the starting sequence of the film provided a slow pace but a high-intensity introduction to the drama. By the time that the students had been introduced to their roles, they were committed to their tasks as researchers.

After the initial meeting with the director, which took place at the front of the classroom and at which the information packs were discussed, an outline of the work provided and the discussion about the graphic sequence took place, the researchers worked in pairs to decide on lines of questioning or specific questions for the roles they were to meet. A very interesting discussion arose about whether it was important to see the mother first and gain her 'feeling' knowledge of the situation or to meet the doctor who would be able to provide hard facts about the situation. Decisions were being made, in fact, about how people learn best and the children were empowered to have direct control over their learning.

The researchers interviewed both mother and doctor by preparing a space at the front of the classroom and arranging themselves in ways which they considered appropriate for the individuals they were to interview. Some discussion took place, for example, as to whether the mother might be intimidated by facing such a large panel of interviewers.

Most of the session's organization arose naturally from the context and significant parts of it were organized by the learning group.

Time devoted to the activity

The session described occupied a 90-minute period but could have taken longer. It was not planned that the lesson should have a specific conclusion since this would curtail the possibility for the learning group to create their own outcomes and reduce potential follow-up activity.

Follow-up activity

A number of potential follow-up activities emerged from the session. The design of the computer graphics sequence which involved the children in considering the basic information about genetic transfer, but for a reason defined by the drama, is an outcome which places the children in a quite different conceptual position towards the information. Instead of being passive receivers they are the disseminators of information and are in a position of power to control it. Time spent on this activity is time spent learning about genetics. Verbal and written reports about the research interviews, photographs, suggestions for further lines of research, and so on, are all potential follow-up material. All these activities take place in an atmosphere of productive tension, provided by the television context of deadlines, economy, taste, and so on. The teacher's task is to continue to use the context to provide this productive tension.

Although no formal assessment was made of the success of this series of strategies in achieving our aims, the teaching team felt that the session was successful in achieving a number of its stated aims. The context did provide fresh demands on the children's language and they responded extremely well to this. They showed very little diffidence in organizing the sequence of events for themselves and did manipulate information they already had as well as beginning to search for new information. The children had no difficulty in accepting their teacher in role. Although it was felt that most of the children were actively involved in the drama, some members of the group contributed very little orally. This concerned the teaching team but seemed likely to be a result of a lack of familiarity with the 'rules' of working.

SUMMARY

We return to our four points again.

1. Which aspects of active learning do they encourage?

Autonomy, personal investment, motivation, sympathy and empathy are all at a premium here. Interestingly, it is not always the class loudmouths who first volunteer into the action. Nor does role play always produce laughter and silliness to cover embarrassment. Pupils can normally respond and enjoy physical action – many of them are well used to working this way in drama and English – and they can relax quite quickly and bring activities to life.

Role play and drama are not to be seen as something special – a big production for which a specialist is needed – but as an active part of science lessons. As they move with and within events in science, the challenge to youngsters (as Baldwin and Wells, 1981, suggest) should be 'Don't tell us – show us!'

2. What planning do they require?

While the discussions that lead to, or occur within, role play and drama can be ad libbed, the occurrence should not be. This is not to curb spontaneity but to prompt forethought and planning. Much of the planning depends upon the aims of the role playing – the combination of possibilities is endless. Players may take on the role of an imaginary person, a real person, or themselves. The situation may be simple or elaborate, familiar or strange. It may be scripted carefully with extensive details of the characters and contexts, or left very much to the players' imagination with only a few sentences drafted for guidance. The time for role playing may be kept short or last for a whole lesson.

There may also be many kinds of learning that take place. It may be first-hand, as in that acquired by participation, or vicariously second-hand, through observation. It may involve learning skills, for example the skill of empathizing or developing concepts, as in case study 24. It may be designed to lead to an increase in sensitivity or a change of attitude. In other words, there are important decisions to be made. These are influenced by the aims of the role play, the level of experience of teachers and students with the technique, and the time available. Thus, topic, group, players, observers, roles, scripts, contexts, cues, clue cards, background reading, timing, discussion of issues learnt, and follow-up work all need deliberation.

3. What organization is needed?

Morry van Ments (1983) has produced an excellent checklist of procedures in the form of a flow chart. It includes such questions as:

1. *Have you set the objectives?* What is the purpose of the role play. Is it to enhance a skill like 'communicating', or change an attitude?
2. *Have you decided how to integrate it within your teaching programme?* Is it a

warm-up session to a piece of work, or a stimulus to get people thinking? Is it a follow-up to a point already being explored, or a means of introducing social or historical contexts? Does it summarize a unit of work, or act as revision which might otherwise have been done in some other way?

3. *Have you decided on the external constraints?* How much time can you allow? Where will you do it? In the laboratory? How much debriefing will pupils need? What materials and information will they require to carry out the roles? Do the major aspects of the roles lie within their experience?

4. *Have you decided on the type and structure of the role play?* How long is it to be? How scripted? What sort of roles? Are there to be observers? What will they be required to do? All of these decisions will affect the way in which the role play is run, debriefed and followed up.

To end this section, we pick up the last of van Ments' questions. He points out that there are several different types of role play, and the choice can affect the running, debriefing and follow-up session. This merits a further quick examination of the possible purposes of role play in science and other structures readers may wish to try for themselves.

Purposes of role play

There may be several. Milroy (1982), for instance, suggests five categories:

1. *Descriptive:* illustrating a process, situation or problem, such as deciding where a chemical plant might be built. Jon Nixon's case study fits well in this category and, in many ways, so does Martin Hollins' example, although his main emphasis was in fact on concept rather than process development.
2. *Demonstrating:* showing a technique, such as describing how to measure or read an ammeter or carry out a scientific process safely.
3. *Practising:* a skill, such as communicating, or listening.
4. *Reflecting:* giving feedback to others on their behaviour as experienced by others, such as the behaviours shown in small-group discussions.
5. *Sensitizing:* increasing awareness of a situation; for example, how animal rights' campaigners feel. Hamish Fyfe's example would come into this category.

Structures of role plays

In two of our examples, the teachers and their colleagues were in particular roles, whereas the pupils had a more open-ended part in the proceedings. Jon Nixon's case study provides an example of the management of role play with a large group of pupils and gives some ideas of how that might be managed in terms of what different groups are doing. Our last section to this chapter goes beyond our case studies in many ways, and describes some of the ways in which different role plays might be managed in classrooms.

There are several mechanisms for actually running a role play. A specialist book on role play will give more detail, but here we lay out two basic methods and explore a few variations.

1. *Fishbowl technique.* This is the most common type of role play. The players play their roles, usually in the centre or at one end of the room, and are observed by others.
2. *Multiple techniques.* This is several role plays conducted in parallel. It is a useful way of working with large groups. It may be several sets of players exploring the same issues, or sets of players exploring different issues around a central theme. Some groups may consist of players, or of players and observers. Three variants which are useful might be:

   (i) *The use of consultant groups.* Each player has a support group who advise and support the role, suggesting how it might be played, what decisions might be made, and so on. The role play can be stopped at convenient points to allow players to consult. Obviously, this is a sophisticated technique which works best in a 'fishbowl' situation.

   (ii) *Role rotation.* This is particularly useful to develop skills. For example, the role of the 'expert' who is demonstrating or explaining can be rotated so that all the class have the opportunity to experience 'expertise'.

   (iii) *Role reversal.* Where issues can be polarized so that two 'protagonists' are involved (smokers and non-smokers, nuclear fuel supporters and non-supporters) they can each be asked to exchange roles. This can help them experience the feelings and explore approaches in the arguments.

4. What else is important?

One aspect which we have mentioned in passing is the notion of debriefing. This is an important idea, for two reasons. First, it should be the time when the major issues arising from the role play are explored. For example, in a role play designed to practise skills, the skills would be discussed and the degree of skill shown by different players explored. This provides the opportunity to talk over both how good they were and how they might have improved. In some cases, the consequences of certain actions can be analysed and reflected upon – it helps students establish their learning. Debriefing is an essential session and adequate time must be allocated to it.

Even more important, it is also the session when players are given time to 'come out of the role'. In some roles, players can feel angry, hurt or confused by the actions of others. Debriefing is essential in giving them time to 'unwind'. That is, to share those feelings with others and talk through them so that they leave the lesson 'as themselves' again, not still with some 'residual role' left.

BIBLIOGRAPHY

Baldwin, J. and Wells, H. (Eds) (1981). *Active Tutorial Work*. Oxford: Basil Blackwell in association with Lancashire County Council.

Bolton, G. (1984). *Drama as Education: An Argument for Placing Drama at the Centre of the Curriculum*. London: Longman.

Dodgson, E. (1984). *Motherland*. London: Heinemann.

Evans, T. (1984). *Drama in English Teaching*. London: Croom Helm.

Milroy, E. (1982). *Role Play: A Practical Guide*. Aberdeen: Aberdeen University Press.

Nixon, J. (Ed.) (1982). *Drama and the Whole Curriculum*. London: Hutchinson.

Nixon, J. (1986). *Drama as a Mode of Learning in the Science Curriculum*. A Workshop Pack. London: SSCR.

Nixon, J. (1987). *Teaching Drama*. London: Macmillan.

Taylor, C. (1987). Dramatic events in science education. *Physics Education* 22, 294–8.

Van Ments, M. (1983). *The Effective Use of Role Play: A Handbook for Teachers and Trainers*. London: Kogan Page.

8: MEDIA AND RESOURCE-BASED LEARNING

INTRODUCTION

This chapter deals in the main with resources for science teaching. As with many resources, even those that are abundant and relatively cheap (in some cases free), it is not just having access to the resources that counts but making the best possible use of them. The case studies in this chapter feature different resources, showing how the contributors employ them to facilitate learning. Media resources and resource-based learning both have the potential to increase youngsters' control of their own learning. Both are underused, perhaps because they require the collation of a range of materials, an activity which can soak up large amounts of time. Time is precious and, when in short supply, many teachers often take refuge in methods they know they can manage with the minimum of preparation and maximum benefit. However, if resources are well used, as described here, what they consume in preparation time can be repaid richly in the classroom. They can bring about quite dramatic changes in the role teachers adopt. More to the point, they can encourage youngsters to control their own learning, enhance existing skills, help create new skills and enable greater understanding. This can free teachers from the hard work of 'up-front' presentation and organization, and allow them to direct, manage and interact on a much more individual level with their pupils.

Our first two contributions deal with the use of educational television. The broadcasts they describe are made specifically for educational use, not for the 'Horizon' kind of evening viewing. This is an important point, because material produced for educational purposes is designed for a specific audience and age range. The language level of the commentary is, for example, carefully matched to the viewing audience, and the content is often drawn from existing syllabuses. The programmes are part of a 'multi-media package', complete with pupil and teacher materials, often with computer software support and sometimes with an accompanying textbook.

Television is a very familiar presence in the homes of students and teachers alike. Despite this, its potential over the years has been very much under-exploited. Perhaps because it is so familiar, it tends to become like the 'box in the corner' – ignored as a means of assisting learning and regarded mainly as a means of entertainment. Perhaps it is this attitude towards the image of 'television as entertainment' to which Gwen Dunn (1980) is referring when she says:

> First and foremost I'd like to see everybody – every single individual concerned with education – accept the fact that television has been invented. . . . It's no good nursery schools, infant schools and those who train teachers to work in them, saying they get enough of that at home and trying to ignore television. Television won't go away. It's time it was given its wedding ring and a crisp veil and married to the respectable body of education.

Of course, not all teachers, primary or secondary, ignore television; many do include it in their lessons.

When asked what purpose television fulfils for them, teachers include the following uses:

- showing experiments that are difficult to do in classrooms;
- showing the social and technological aspects of science;
- bringing 'field study' materials cheaply into the classroom (Bentley and Watts, 1986)

It can achieve much more than this, of course. As we have pointed out elsewhere (Watts and Bentley, 1987) youngsters learn a great deal from television. Quite what they learn is a different matter. Their comments indicate that while many science education programmes are fairly uninteresting and irrelevant, they are a welcome break in an even more boring classroom routine. A great many science programmes which they see are unexciting: though, thankfully, not always – some images are provocative and imposing. The pace, language and density of information is not always 'viewer-friendly', but some programmes make special efforts to be interactive and maintain a pace which, while being fast enough to be interesting, is still slow enough to help understanding.

For us, one function of active learning in school science is to wean youngsters from overt dependence, towards self-direction and independent learning. Educational television is a useful contributor to such a function. Despite the somewhat negative views from students which we noted above, educational television can have a variety of roles in lessons.

It can be an entertaining 're-vision' of ideas previously raised in class; a challenging counter to youngsters' own intuitive ideas; a source of powerful visual images; a distinctive means of explaining complex arrangements of ideas or apparatus; and a way for youngsters to direct their own learning of important issues. The majority of youngsters are very familiar with television and video-playback machines, and with their mechanics of access and operation, and are at ease in using them as sources of many kinds of information. Such expertise could well be better exploited in ensuring that programmes are made with students in mind as the 'controllers' of the playback button rather than their teachers.

We referred earlier to the attitudes which many teachers hold towards television. Some of this may indicate a reluctance to use television, because in its normal everyday domestic use, it is generally considered a very passive medium – the greatest interaction with programmes being to change channel or to switch off. Thus the pervasive ethos of television generally, is that the viewer is passively receiving information and there is little active learning involved.

This need not necessarily be the case. The substance of our first two case studies reflects rather a different picture. Our contributors are concerned with the contribution that television programmes can make to developing youngsters' understandings of ideas in science. They display quite clearly how this might be done in an interactive way. Robin Moss's contribution explores some of the ways he has found of helping youngsters to do this, and Philip Munson's case study provides a further example of the ways in which youngsters can use television in a science lesson to enhance what they know and gain greater insight into scientific ideas.

Teachers' and pupils' attitudes, as we mentioned briefly earlier, may be a major stumbling block in the use of television (apart from the daunting logistics of gaining access to a set plus video-recorder, in the right place at the right time, with the right programme). Few see educational television programmes as a serious medium by which science can be learnt. At best TV is treated as support material for other classroom activities, frequently as a break in routine, often simply as an entertaining interlude. Both the information base and the visual images presented are considered to be ephemeral – the effect on the learner is transitory, soon to be forgotten. Where programmes are used at their most intensive they are still reinforced with written material – often the student is required to make notes or complete follow-up worksheets in order to establish the information that has been provided. It cannot be learnt, it seems, without being written down. It cannot be revisited or revised without being the personal written property of the learner: visual and aural learning of this kind is not normally institutionalized by being tested or assessed. Where programmes are used at their most frivolous, television and video are seen merely as entertainment and diversion.

This view is not universal, of course. In our research (Bentley and Watts, 1986), youngsters rated visual images very highly as a mode of learning, despite their observations that, too often, the images of science and scientists are uninspiring and dismissive of women. Images, they said, are a means of learning. They help ideas, principles and issues to 'be seen'. That is, to be interpreted and internalized in a meaningful way. For many youngsters, visual images are an important means of reaching an understanding of complex matters. Moreover, some of the images provided on screen are strong and can have a lasting effect.

If educational television is to be used to its full potential, then teachers need to be sure what that potential is, and how to make the best use of it. They need to be critically aware of the strengths and weaknesses of the programmes they use in their teaching and plan so that these enhance the scientific ideas they want youngsters to explore. Television can be a passive medium, or it can be a very challenging one. It can, for instance, be a very formative influence on both the science learned and the images of science and of scientists that are presented. It can introduce pupils to new and unexpected phenomena and stimulate them to activity. Such stimulation does, however, presuppose that programmes have been made with that kind of use in mind – to be 'user-friendly' for the learner, interactive with other media, geared to challenging ideas, structured for ease of reviewing and revision, relevant to youngsters, and human in approach. As Peter Bratt (1986) says: 'In short, television can add a whole new dimension to the teaching of science in schools.'

Rod Dicker's contribution is somewhat different. It too focuses on resources, but on the ways in which teachers can use industry as a source of information and context for their science teaching. Again this seems to be a much underused resource. As Rod points out, industrially produced materials are often free and very much more up-to-date than most textbooks. The kind of industrial links which can be established between schools and industries are an important way of providing that social, technological and economic relevance required by many GCSE criteria. The materials can lead further than this. In the *Times Educational Supplement* of 27 February 1988, for instance, the relationship between Austin Rover and local schools seems to be a symbiotic one. Apprentices at Austin Rover had the opportunity to work with GCSE pupils, explaining points, assisting them in learning new skills in the classroom context, and thereby enhancing their own understanding. Pupils had the opportunity to learn for themselves some of the applications of their work in school and develop a better understanding of principles in action.

CASE STUDY 26: USING SCIENCE EDUCATION BROADCASTING AS A PART OF THE SCIENCE CURRICULUM

Phil Munson

Introduction

There is a vast variety of high-quality material available from radio and television which can be used to enhance science teaching for many pupils. However, much of this material is under-exploited by many teachers. There would appear to be several reasons for this. Three main ones, which I shall address here, are:

- difficulty in determining in sufficient detail exactly what is available and matching it to curriculum needs;
- lack of time and experience in developing appropriate teaching strategies which will ensure its effective use as a learning medium; and
- problems associated with ease of access.

These problems, and possible solutions to them, are explored more fully in the following example, which draws on experiences in presenting scientific processes of investigation/problem solving to a class of 12-year-olds.

Prior planning

The first step in planning is the actual decision, best taken at science department level, to include aspects of broadcasting in the lessons and teaching approaches. The decision is best taken at this level initially because access to suitable material, and more important, its availability, can be guaranteed.

The second level of planning involves what should be included. Obviously content will be an important consideration here, but other aspects of the programme are equally important. In using broadcast material in a unit of work on problem solving and investigation, it is important to have programmes which encourage this approach in an active way. Programmes which present a didactic image of science are of little value. Fortunately, approaches such as problem solving have been recognized by both the BBC and the Independent Television companies, and they are now producing a great deal of relevant and suitable material for both television and radio. Many of the programmes I viewed and elected to use, consist of short 'magazine' type items, specifically designed to pose questions and stimulate investigations. A common format of such programmes is to have some kind of 'break' between the sequences – a logo or tune, or even a fade to black. I have found this very useful because it means I can use the short sequence as a stimulator which the pupils use to frame their investigations. It is rather like doing a demonstration to spark something off. It only lasts 4–5 minutes and yet it is the basis for a whole lesson's problem-based work.

Once I had decided from the programme booklets which series or programmes – or in some cases, parts of programmes – to use, I then had to check them for their suitability. This meant recording the programmes, a time-consuming and often hazardous business! Finally I had to view them to ensure that they would meet my needs and that they did what I intended them to do. I also found it very useful at this stage to acquire the written and computer support which went with the programmes. Although this is a very time-consuming exercise, it is an essential pre-requisite for the successful exploitation of the medium, and it does provide dividends later in terms of the enhanced learning and motivation of the students.

Teaching use radio and television programmes

The experiences outlined in this section refer to material which had been pre-recorded, i.e. not as it was being transmitted. Pre-recording is important because it solves many logistical problems, i.e. being in the right place, with the right equipment, at the right time and for the right length of time. It does, however, mean that you have to get yourself organized well in advance!

Traditionally, educational broadcasting has been a rather passive medium. The usual television format has been a 20-minute programme which has presented information in a sometimes interesting and occasionally entertaining manner, exploiting the professionalism of the presenters. Pupil interaction has been minimal. Often they have been used as summaries at the end of a topic, or as a means of revision. More recently, the format has changed. The individual items which make up the programme are short, sometimes only 3–5 minutes long. I have used such individual items in a variety of ways:

- as introductions to a lesson, to set the topic up in a different way, bringing in examples from industry, or parts of the animal or physical world that I have little easy access to;
- as a short 'trailer' to stimulate discussion;
- as a means of introducing a 'brainstorming' session. For example, showing a phenomenon, then switching off and asking pupils to

brainstorm what they think might have caused the phenomenon, or explain what they think is happening and why;
- to set the scene for a practical investigation.

Regarding practical investigation, one series of programmes – *Scientific Eye* – provided a good context for the work we were doing on solving problems associated with heat. The programme was designed to help pupils understand the processes of science, and deliberately drew their attention to skills such as hypothesizing, testing out an hypothesis, drawing conclusions and re-designing. In the particular programme I used, the problem set on screen was to investigate the conditions under which a cup of tea would cool down fastest. I showed the setting up of the problem, which involved some sequences about heating and cooling, and then switched off at the point at which the pupils on film had been asked to take part in a 'cooling race' for their investigations. I set my pupils the same problem, asking them first, in small groups, to brainstorm what solutions they might try, select one which was feasible and suitable, and carry it out. We compared our results and solutions as a whole class, and then I showed the few minutes of the programme that remained at the end of the lesson. This showed the pupils in the film trying out their solutions, with their results. My class of 12-year-olds were able to use the experience to both compare their own solutions – many, inevitably, were the same – criticize the techniques of the youngsters in the film, and also learn from their methods of handling the problem. This raised many discussion issues about scientific processes and was very valuable.

While television is a useful and stimulating aid for teachers, radio can be equally as challenging. In fact the radio programmes I have used, which are designed to encourage problem solving, have a distinct organizational advantage over television, since the hardware for radio is easier to access and operate, especially since the advent of 'personal stereo'. I have found them particularly useful when organizing problem solving in small groups. The group listens to a storyline on the tape, which sets the scene for a problem-solving exercise. There are further extensions and information available on the tape for after the problem has been completed, and the availability of tape-recorders and headphones has meant that several groups can work on different problems in the room at the same time. The series is supported by written materials with apparatus lists, pupil sheets, ideas to help low-achieving students and extension work for those who need it.

For some topics, where the issues are based more on the societal implications of science, one way of working is to make the video or the radio programme the focus of the final part of the lesson. Groups are asked first of all to produce a script around the topic featured in the programme. They are asked to outline, research and script all the aspects they would want to have in their production. This may be an exercise that takes one or more lessons at the end of a topic, so that students can draw on expertise they have already acquired during their previous lessons. Or it may be an ongoing homework exercise which takes place over a series of weeks, and involves pupils in collaborative tasks for homework, researching and changing their script as they learn new things in the lessons. The final denouement, after scripts have been read out – or in the case of some students, perhaps sold on the problem-solving radio programmes, recorded onto tape at home – is to show/listen to the actual commercially produced programme. It is interesting to compare what pupils would include with what professional producers put in their programmes, and to look at the way youngsters have of explaining complex scientific ideas to their peers in comparison to the ways producers choose.

Organizing the use of radio and video

It is all very well to extol the virtues of educational broadcasting as a learning and teaching medium, but many teachers, although very willing to exploit it, find that the organizational difficulties are so overwhelming that it is often not a viable proposition. To begin to solve this problem it is necessary to return to a point made in the introduction, that is, that the inclusion of educational broadcasting in the science curriculum must be a decision that is made at science department policy level. To exploit video in the ways that are outlined, it must be available in the laboratory. The tapes must be stored in such a way that pupils and teachers can access them easily and obtain items on them with relative simplicity. If videotapes are to fulfil a function in enabling pupils to become active learners, then, like books, they must be a resource that the pupils can access, use, revisit and learn from with ease. This has implications for storage and cataloguing procedures. Ideally, videotapes ought to be able to be used for pupils to look up information, revise for examinations, access up-to-date social and technologically relevant issues of science for GCSE course work and watch for general scientific interest. The science department must decide how these potentials can best be met within the geographical constraints which are inevitably different for every school. Unless these fundamental requirements are thought through and implemented, it is probably very difficult, but not impossible, for an individual teacher to use such resources to their full potential.

Given some of the constraints, what do teachers get out of educational broadcasting in the long run? The quality of the product is such that it is well worth spending the time opening up the learning opportunities to your pupils. They lend themselves, with support, easily to pupils at any level of achievement, they are one mechanism for assisting bilingual learners to better understand science, they exploit visual learning and they are extremely motivating and stimulating.

CASE STUDY 27: VISION AND SCIENCE LEARNING EXPERIENCES

Robin Moss

I don't remember a great deal from my own study of science at school. In fact, if I'm honest about it, I remember only two things: Otto, the aptly-named vacuum pump, being cranked round and round to evacuate the air from a large 'empty' can, until the atmospheric pressure crushed it; and a satisfying experiment involving a test-tube half-full of spit and breadcrumbs, to demonstrate the action of saliva on starch. I had won a place at a good independent school, through the '11+' examination and I am sure 'old Wilkie' (he must have been all of 35) was a dedicated and effective general science teacher. Yet my science schooling, albeit cut short by specialization as a classicist at the age of 13, was almost totally useless. That's not quite fair. We were all drilled in the 'scientific method' and I could recite a version of it, if required, but really, apart from that and a memory of Wilkie's face hovering above Otto, I can recall little.

A dire tale, but not a rare one, I fear. It has left me curious about science and scientists, and somewhat humble, too. *The Ascent of Man*, for example, filled me with excitement and enthusiasm, as it did so many, and I have often quoted Bronowski's words at the end of that remarkable series of television programmes:

> We must not perish by the distance between people and government, between people and power, by which Babylon and Egypt and Rome failed. And that distance can only be closed if knowledge sits in the homes and heads of ordinary people with no ambition to control others. (Bronowski, 1973, p. 435)

Bronowski was speaking of 'the democracy of the intellect', the need for education to be effective so that citizens can understand their own society and so govern it. He was talking about education and its link to power. The failure of my own scientific education left me a weaker person, Bronowski would have rightly said, and thus less able to act as a citizen.

The Royal Society wholeheartedly agrees. In 1985 the most prestigious body of scientists in Britain, if not the world, launched a new campaign aimed to improve the general understanding of science, a campaign specifically aimed at broadcasting. Bronowski, of course, was delivering his ringing declaration on television, at the beginning of the previous decade. Television is often described as 'a medium of great power' by politicians, usually when they are lamenting its alleged failings. Less attention has been paid to the power of audio-visual resources in education, particularly in schools.

At about the same time that Bronowski was at work on *The Ascent of Man*, I was a supply teacher in a secondary modern school in North-ampton. Given (what else do you give to supply teachers?) the low-ability early leavers to 'manage' during a humid summer term, I turned in desperation to the cupboard of the local Teachers' Centre. There I found an under-used single-camera video kit.

Armed with this I could offer 4S something else. They reacted well to the invitation to 'make a telly programme'. We created production teams and allotted roles. Pupils wrote scripts, arranged running orders, argued, shared ideas and discovered new skills. What was interesting, on reflection, was the speed with which the children picked up new technical skills. They also showed qualities none of us (not even the other pupils) knew they had. In 1976 a Schools Council project headed by Carol Lorac and Michael Weiss explored the same territory and reported on a number of classes at work on making their own audio-visual material in *Communication and Social Skills* (1981). Here is one example of what happened, with a group of low-ability fourth year boys in a school in the North-East, working on a rural studies tape-slide project 'Your Daily Pinta':

> The intention of the programme was to show how milk is produced 'from the udder to the doorstep'. What is of greatest interest is the way in which the pupils organised their thinking in planning the programme . . . the pupils went through two distinct planning stages . . .
>
> . . . even the least motivated children responded. Pupils who had not been significant before suddenly became significant because every single cog was necessary for the machine to function. . . . The sheer quantity of purposeful talk increased dramatically. . . . The content of the learning was so close to the children. It was their school, their film, their cow, their visit, not someone else's book or film! (Lorac and Weiss, 1981, 56–7)

The Project's authors define the method as involving pupils in responding to a topic as a group and expressing their findings audio-visually. They concluded that the process enhanced pupils' language skills, and increased learning generally in such areas as encouraging exploration of the connections between theoretical ideas and the practical world. Clearly, with present exhortations to science teachers to create situations that help pupils apply their science, using audio-visual equipment has some real benefits.

It is often claimed that educational television plays a part in helping pupils with this linkage. Indeed some of the earliest ITV Schools broadcasts in 1957, the first seen in Britain, paid particular attention to showing the application of scientific theory to contemporary society. Lorac and Weiss were talking about something more. They were discussing the liberating experience of combining to plan a logical approach, solving practical problems, and combining to understand and then express ideas in audio-visual terms.

To some extent the development of the last few years in science teaching have incorporated many of the ideas covered in *Communication and Social Skills*. Have they given enough consideration to the power of the eyes? Psychologists tell us that our eyes bring much more information to the brain than any other organ. Modern primary schools take full account of this, as a glance (for eyes are swift as well as powerful) will show. Studying the Interim Report of the Science Working Group on the National Curriculum (1987), I am particularly encouraged by passages on secondary school learning that recommend a larger place for vision in science education. In choosing a learning experience for science at primary and secondary level, the Working Group recommends, among other criteria, the following list:

Will the experience

- stimulate curiosity?
- give opportunity for developing . . . skills?
- give opportunity for developing . . . curiosity and co-operative working?
- give opportunity for developing basic science concepts?
- help children to understand the world through their own mental and physical interaction with it?
- involve resources and strategies accessible to all teachers?

For all these (and indeed, to a lesser extent, for the other half of the list) broadcast television can offer invaluable stimulus and support to science learning activities in and out of the classroom. What other relevant stimulating resources are currently made available to every school in the UK free of charge? The BBC and ITV companies now have 30 years of production experience, and frequent staff contact with children and science teachers on which to draw.

Even the harshest teacher-critics of schools television (and there are still some) admit we sometimes get it right. They tend not to agree about which resources are right for them, but they will admit we try to deliver our broadcasts *as* resources for use, stimulating opportunities to develop skills in science and technology.

But what has this to do with Otto? Remember my old vacuum pump friend? A television programme about Otto would not have stuck in my mind for 30 years (although Bronowski will), and cranking the pump myself might have been an even more evocative experience. The point I would leave as a question for science teachers is as follows: If science learning experiences, however stimulating and wondrous, still converge upon *written* (and drawn) personal reports, will children who find writing and drawing difficult lose out?

The excitement of *my* experience as a teacher in Northampton, as replicated in the Lorac/Weiss project, was to see children who did *not* shine in conventional terms, understanding and expressing ideas effectively for the first time, by doing so in visual terms. Naturally, I believe video material has a vital place in the science curriculum. Should not the creation by children of video on science subjects also have a place?

CASE STUDY 28: THE CLASSROOM USE OF INDUSTRIAL RESOURCE MATERIAL

Rod Dicker

Why use industrial resource material?

Recent changes in the school curriculum, such as the introduction of GCSE and TVEI, have resulted in an increased need for resource material that can be used within the classroom. Many teachers like to assist their pupils to understand the relevance of the science that they are studying; they may wish to provide a varied programme of learning opportunities to aid motivation and interest; they may wish to use resource material that is as up-to-date as possible. Textbooks are obviously written to meet the educational or training needs of a particular group of pupils, or an examination syllabus. However, textbooks can sometimes prove to be an unsatisfactory means of providing information. For example, they can provide information that is somewhat out-of-date, reflecting processes and techniques that are rarely used within the present industrial setting. Textbooks can also be an inaffordable luxury, the budgetary demands of a school department may necessitate the alternative purchase of equipment and consumable materials.

Industry is increasingly organizing various types of liaison with educational establishments such as schools and colleges. The production of written and visual material will often be produced as part of this liaison process. Sometimes this material is produced to support a visit by an industrialist when visiting a school. More often, industrial resource materials (IRM) are produced to publicize the work, or products, of an industry, to provide a means of replying to requests for information from school and college pupils, to illustrate industrial applications of science, to support project work in schools, and to change the attitudes of teachers and pupils. A wide variety of IRM is produced including wallcharts, booklets, sample packs, films and video-cassettes, slide/tape presentations, and resource packs consisting of a number of resource items relating to a particular theme. There are other types of resource material that can be used by pupils and teachers. This is material produced by the industry for internal use, but can often be obtained and used as a source of information.

How to find suitable industrial resource material

Industrial organizations that genuinely wish to help education often go to considerable effort to assist teachers in becoming aware of the existence of the nature of the resource material that they have produced. They will often publicize their material through educational journals and papers such as the *Times Educational Supplement*, and they also send literature and catalogues directly to schools. Many teachers, however, fail to become aware of the wide range of materials that can be obtained. Very often, the plethora of paper and information that is sent to a teacher results in much of it being put into the waste paper bin with only a cursory glance. Alternatively, the information may be neatly stored in a filing cabinet, but with the same result! Teachers that I have spoken to have described the process of finding IRM as a matter of 'sheer luck', or a 'little random'.

Teacher and Resource Centres often hold a stock of examples of various types of IRM which can be consulted, and sometimes borrowed. Since many teachers are extremely busy people with little time to spend searching for different items of resource material, another effective means of finding suitable IRM is by talking to teachers from other establishments at meetings and conferences. Much can be gained from the experience of other teachers.

A more recently introduced method of discovering suitable resource material is through the use of computerized resource databases such as NERIS which can be searched using a BBC computer and telephone modem. Computerized databases which can be accessed through a telephone modem can save a considerable amount of the teacher's time – if the hardware and telephone line are available for use! Searching databases in this way can be expensive, and therefore restrictive, if the search hasn't been organized prior to logging into the system. Considerable time can be saved if the information is rapidly accessed and then downloaded onto disc rather than onto the computer's screen and then printed. Rapid downloading onto disc keeps the school's telephone charges low, and it allows a leisurely and repeated search of the information obtained from the database. If the school or local education authority's administration system allows, certain suppliers of resource material can be contacted through the database and copies can be ordered.

Another database of industrial resource material that exists is held by the UBI (Understanding British Industry). This organization holds many different examples of industrial resource material, reports, etc. It is also able to provide a search facility for people like teachers who wish to locate material for use in the teaching of a particular topic or subject. Other organizations, such as SATROs, SILOs and SCIPs, can also provide help in finding suitable IRM.

A considerable amount of IRM can be provided free of charge to educational establishments. However, certain producers will impose a moderate charge, not to meet the production costs, but often to prevent the 'magpie' syndrome where teachers will collect, but not use, any free resource item. The imposition of a charge helps to reduce the demand made by teachers for material and might, therefore, help industry to provide a wider range of resource material.

How to choose suitable industrial resource material

A number of factors need to be taken into consideration when deciding whether to use resource material produced by industry. Often teachers would like to obtain class sets of printed material, rather than individual or a small number of copies. However, this need not be an overriding factor in the decision-making process. As with the selection of any material that a teacher wishes to use in his or her classroom, there is a considerable element of personal choice. Every teacher develops a particular way in which they present information to, and encourage learning in, their pupils. Besides these personal factors, there are many features commonly used by teachers. One of the most important of these is the relevance of the material to the curriculum. However, there is no reason why IRM need only be used for 'normal' class activities – it can be used as extension material for the more able pupil.

Relevance to the curriculum should not be the only consideration. Resource material can only be suitable for direct use by pupils if the level of the language and the technical level are suitable for the target group. Much material can lose its potential effectiveness if either of these criteria is not pitched at the correct level: neither should be too difficult or too easy. Material that has been produced for a group or category of reader that has reached a higher level than the group or class that a teacher wishes to use that material with, will be left having seen or read something with which they cannot fully relate. However, even if IRM can be rejected on these latter criteria, it could be suitable as a source of background information for the teacher. Also, parts of the resource material might be suitable for use with modification by the teacher, e.g. as worksheets, posters, etc. Visual presentation of IRM is another important consideration. Visually attractive material can markedly affect the usefulness of the material.

Certain types of material show a particular bias, some of which might be regarded as political. Obviously, in this situation, the teacher is obliged to achieve a balance by presenting the alternative view. This may involve finding industrial resource material from another source, or presenting the alternative oneself. Such bias in industrial resource material might be found when an organization is intent on advertising and convincing the reader of the merits of a particular industrial process, such as the use of nuclear power. In the case of nuclear power, a balance might be achieved by selecting resource material from a conservation group such as 'Friends of the Earth'.

The ability to present information in such a way that it is appealing to pupils, is informative, understandable, relevant and involves pupils in suitable activities, is one that usually requires training. Much material that can be used within school has not been written by a practising teacher. However, a combination of the skills and knowledge of teachers with those from industry should result in the production of suitable material for the classroom.

It should be remembered that resource material should not merely provide written or visual information. It should be able to provide some of the learning needs of a particular group of young people. Teachers and pupils have individual needs. Resource material should be designed to allow flexibility of use without neglecting the various scientific principles and concepts for which it is being used.

Where to use industrial resource material

At the present time, marked changes are taking place in the courses being provided by schools and colleges. These changes have resulted in the introduction of the Certificate of Pre-vocational Education (CPVE), Technical and Vocational Extension (TVE), and the General Certificate in Secondary Education (GCSE).

Often, industrial resource material can contain up-to-date information – textbooks are very rarely able to satisfy this criterion. IRM, for example, in the form of current bulletins or news-sheets, can be used in the classroom, library, or resource centre by pupils undertaking project or assignment work.

Together with other kinds of support from industry, such as visits and periods of work experience, industry is able to provide a considerable service to teachers and pupils. However, it is only by teachers talking about their experiences in using the various resources that industry provides, that we will find a greater, relevant use of IRM in schools and colleges. The use of suitable resource materials can show the relevance and applications of the topic as well as making them more aware of industry as a source of employment. In addition, the use of resource materials within a suitably structured curriculum and pedagogic system could well promote additional, desirable traits. Through the use of resource-based and independent learning, it has been found that pupils become more self-reliant and independent, questioning, stimulating, pleasant and happy.

SUMMARY

This chapter has covered a variety of possible resources for science education, with an indication of how they can be obtained (often at little cost), and ways of using them to their best advantage. However, our interest in the use of a wide range of resources is only pertinent if it contributes to active learning. We examine that question first. But rather than continue by looking in detail at the planning and organization required, we present this last chapter a little differently; first, because many of the planning and organizational details have been covered by the contributors in their case studies and, secondly, because we want to focus mainly on the choice of resource. Matching materials to the curriculum is an important task that deserves full consideration. In practice, it is usually a rather agnostic and *ad hoc* process that undervalues careful deliberation. As part of our summary, then, we present a series of criteria by which resources might be selected.

Which aspects of active learning do they encourage?

Throughout this section we take the word 'resources' to mean things like posters, pamphlets, books, videos, film loops, tape-recordings, educational broadcasts, photographic slides, film strips, library films, photographs, etc. While we have, in earlier chapters, used the term 'teacher-as-resource', we exclude teachers in this context and consider them as managers, and focus instead on inanimate resources.

The answer to our question clearly hinges on the use to be made of such artefacts. Of themselves, resources cannot create active learning. They can be constructed carefully so that they are as interactive and user-friendly as possible. They can be both syllabus-related and student-relevant. They can reflect positive images of people and cultures and be couched in language that is directed at particular groups of students. It is possible to develop all of these strands during the production stage. None of these aspects, however, alone or together, can bring about active learning. That needs active learners, and teaching for active learning.

The latter clearly involves the ways in which teachers choose to employ materials and help students to understand and make use of them. Our case studies show examples of how the crafted use of resources can encourage youngsters to take responsibility for their own learning, and still continue to be in control as it develops. That is, materials can be relevant and important aspects of the students' study of the scientific world. They can be helped in the tasks of solving problems and making decisions and, as in case study 27, students are able to display their understanding of science through the making of a film. This resulted in a tremendous sense of well-being, an enthusiasm matched in case study 26 where resources provide the all important stimulus for active learning to occur.

What else is important?

Anyone who has shopped around for materials at a big book exhibition – or, more particularly, at an ASE conference – will know what 'information overload' means. Rod Dicker points to the vast array of materials that are available – so how can we make sensible choices? No set of materials can be perfect for all purposes: we all have different needs and an eye for different bargains. What we cannot afford to do is waste either time or money – resources can so often lie on the shelf unused. Neither must we see them just as 'fill in' materials for quick finishers, slow learners or cover lessons. Nor can we afford to make unprincipled choices, to pick randomly or capriciously through a book catalogue or list of commercial promotions.

But what kind of principles? Here we set out some criteria for choosing good science education resources. We present them in the form of a checklist which considers the following questions:

- How can resources meet the projected needs of both teachers and learners?
- How well are they geared to the current considerations of the ways in which students learn science?
- Are the resources stimulating and motivating?

- Are they compatible with our general aims and objectives for science education?

We start from the premise that resources, such as those described by Rod Dicker and Robin Moss, have very particular roles in school science. We see these as being:

- to motivate and stimulate pupils' ideas in the classroom by both reflecting pupils' own experiences and providing a wide range of experience not easily or readily accessible to them;
- to illustrate the cultural and technological contexts of science and provide some insights into the ways in which scientific solutions have different relevance to different cultures;
- to present pupils with conceptual frameworks different to their own so that they might see some point and purpose in moving towards a greater understanding of formal science;
- to develop images of science and science education that highlight a variety of methods and approaches;
- to present images of science and science education that dispel present stereotypic views of women in science, and science as a monocultural activity.

These points are ones that begin to shape a framework against which resources and their use in science classrooms might be considered. The criteria broach five main areas of concern which we have explored elsewhere (Bentley and Watts, 1986) – Presentation, Plasticity, Utility, Substance and Image.

Presentation

By this we mean the visual and aural effects the producers have attempted to create. It takes into account aspects such as animation, use of diagrams, use of appropriate filming techniques and techniques for stimulating and maintaining interest in text and film. It examines, for example in the case of television and radio, the match between the expectations created by other forms of media presentation, and the production of educational resources. It looks for presentation that is in keeping with other models of production that youngsters see elsewhere. The questions to be asked about the resources are whether they:

- use methods of presentation that are modern and in keeping with what youngsters see in other presentations?
- include sequences of animation, diagrams and, where appropriate, simulation, which are clear and enable the understanding of the science content?

- present an available language level for the producers' intended audience?

Plasticity

There is a clear need for resources to be as interactive as possible, preferably to be able to be used as part of a mixed package (e.g. including video, text and software) and to provide facilities that cannot easily be achieved in the classroom or on field trips, and which allows for the maximum flexibility of use. This may mean the ability to be used by either teachers and pupils together, or either group independently of the other. It may mean being able to use part of a resource, such as a computer program or a video, for a particular purpose, without assumptions about prior knowledge confounding the issue. Therefore, do the resources:

- have the capacity to be interactive with the intended audience; do they, for example, provide stimulating questions and challenges, set problems in a context relevant to pupils?
- have the capacity to fit easily, without major revision, with other types of resources such as video, radio, slides, texts?
- provide 'extended laboratory' work?
- encourage autonomous, independent learning?
- capitalize on those areas where resources provide more than a teacher can do?
- relate theoretical issues to previous understanding, in terms of a youngster's view of the world?
- provide clear statements for teachers and users about prior knowledge expected by the levels of understanding utilized in the materials?
- use investigatory and problem-solving approaches to science?

Utility

The content (and methods) of science being depicted must be relevant to the syllabus. That is, if resources are to be syllabus-orientated, they must be up-to-date. However, there are other kinds of utility. They might also show science to be useful in both societal contexts and personal terms: resources can be a source of information for both examinations and life outside school. That said, do the resources:

- illustrate both personally and societally relevant applications?
- fit with the syllabus?
- contextualize the issues raised, i.e. give some idea of the applicability and point to the scientific investigation?
- present a personalized approach to science, i.e. encourage youngsters to appreciate that their own views of science can be valid?

Substance

The content of resources must be contemporary, imaginative and crea-tive. Ideally, they should deal with issues in ways which present theories at both general and specific levels, and which encourage pupils to take initiatives in terms of their own ideas in science. They should illustrate the technical, environmental and historical applications of science. In choosing resources, then, do they:

- provide some information base?
- present contemporary content?
- provide examples of different methodologies in science?
- portray historical aspects of science?
- present technological issues and applications of science?
- present environmental issues?
- operate at both a general and a specific topic level?

Images

Here we deal with images of science, scientists and science education. Again, ideally, these should be positive images in that the text or television programmes present 'good practice'. For example, they should be chosen to be positive about the needs of girls and of pupils from different cultural backgrounds, of relevance to youth culture and interests, and promote positive views of scientific practice. Science must be plausible and intelligible (Strike and Posner, 1985) in that it must be believable, i.e. acceptable in terms of its theoretical development, language level, assumptions of prior knowledge, etc. Can we say, then, that our resources:

- provide positive role models for girls?
- provide positive role models for other cultures?
- promote an interest in creative, imaginative science?
- portray science as something that is constructed by individuals within a society?
- include some discussion of failed experiments, to show that scientists are fallible?
- portray science as the problematic collection of evidence towards theory?
- interweave between school and industrial experience and so relate to other aspects of science and school work?

Obviously, it would not be possible for any one resource to meet all of these criteria. However, those that meet most would presumably be more easily justified than others. An assessment of which of the criteria were *not* met would allow one to decide whether or not to use the resource at all and, perhaps, to see what other resources had also to be provided to make up the deficiencies. However, we would stress that the selection of resources to meet the criteria above is only the first step. What is more important is how teachers choose to use them. Our case studies give some clues to this.

BIBLIOGRAPHY

Bentley, D. and Watts, D. M. (1986). *Looking for Learning: Some Perspectives of Science Education Television.* London: Independent Broadcasting Authority.

Bratt, P. (1986). Good practice in using educational television: A producer's viewpoint. In *Review Number Seven: The Newsletter of the Secondary Science Curriculum Review* (Ed. A. Larter), Spring. London: SSCR.

Bronowski, J. (1973). *The Ascent of Man.* London: BBC. Department of Education and Science (1987). *National Curriculum: Science Working Group. Interim Report.* London: DES.

Dunn, G. (1980). Television and the education of the young. *Journal of Educational Television* 6 (2), 47–52.

Lorac, C. and Weiss, M. (1981). *Communication and Social Skills.* Wheaton.

Moss, R. (1983). *Video: The Educational Challenge.* London: Croom Helm.

Strike, K. A. and Posner, G. J. (1985). A conceptual change view of learning and understanding. In *Cognitive Structure and Conceptual Change.* (Eds L. H. T. West and A. L. Pines). London: Academic Press.

Times Educational Supplement (1988). The apprentices who learn by teaching. 27 February.

Watts, D. M. and Bentley, D. (1987). What makes good science education television? *Journal of Educational Television* 13 (3), 201–16.

9: SUMMARY AND DISCUSSION

Mike Watts and Di Bentley

Like many Latin abstract terms, the origin of the noble concept of 'education' is to be tracked down to the agricultural vocabulary of the farmyard (v: L. R. Palmer's The Latin Language, *passim*). It is indeed related to edere – to eat, but is derived from its variant educare – to cram, to stuff full, applied to geese and the like. (Letter to *The Guardian*, 1980)

INTRODUCTION

In Chapter 1 we began to put together a framework for active learning. In each chapter we have tried to fill out the frame and show how active learning might be developed in the science classroom. We intend the book to be useful to a range of people, from those who wish simply to try out new ideas to those who are looking for a whole new way of working. While reading itself can be a form of active learning, reading this book alone would be only a small part of putting ideas into practice. Hopefully there is enough provided in the case studies and the bibliographies to interest those who want to pick-and-choose their way through some alternatives to enhance their usual classroom work. Some groups of teachers have already begun the task of revamping the way they work [e.g. the 'Alternative Teaching and Learning Strategies and Science Education' (ATLAS) group in Surrey].

We noted earlier that many of the classroom activities we have discussed deserve a book to themselves and the bibliographies indicate where this is already the case. What we have tried to do is arrange the case studies to provide a range of 'real-life cameos' side-by-side so that teachers can compare alternatives. We hope it is clear that the chapters do not contain infallible recipes so much as suggest a menu for good practice.

However, just as all menus have a common theme determined by the tastes and interests of the chef, so we see several common themes which run across the case studies and chapters. This chapter concerns a number of contemporary themes that thread through teaching and learning in schools. They are not particular to science but, since our interest lies in school science, we use that as our vehicle. We tackle our themes through a series of questions:

- How do we choose between the different teaching strategies?
- How can we come to terms with active learning?
- Is one teaching method better than another?
- How can we gauge the success of a particular teaching method?
- What of teaching for equal opportunities?
- Where to next?

As we write, the implications in the small print of the National Curriculum can be glimpsed more clearly. One general fear is that a heavy emphasis on the content of science will inevitably mean that any diversity in teaching methods will suffer and be diminished. Teachers will teach to the attainment targets and neglect variety. That is, there will be no incentive to innovate if teachers 'have to get through the syllabus' – a syllabus that lasts from 5 to 16. However, from the rubric so far, there would seem to be no reason to fear an unavoidable retrenchment of classroom styles to traditional didactic teaching. Whether that happens or not will be open to many influences – curriculum policies, methods of assessment, teacher appraisal, contracts and conditions of service, accountability at school and local level among others. To pick up on our themes we begin with the question: How do we choose between the different teaching strategies?

In their interim report the National Curriculum Science Working Party (DES, 1987) have this to say:

> The way science teaching is organised in secondary schools can vary and may include courses in the separate science subjects as well as courses organised around topics or themes. Whatever type of course structure is adopted, however, selections of learning activities need to be made. At both the primary and secondary level the following criteria should govern the selection of learning experiences.
>
> Will the experience:
> - stimulate curiosity?
> - give the opportunity for developing scientific and technological skills?
> - give opportunity for developing attitudes relating to scientific and technological activity including curiosity and co-operative learning?

- give opportunity for developing basic science concepts?
- relate to the interests of children at a particular age and to their everyday experiences?
- appeal to both boys and girls and to those of all cultural backgrounds?
- help children understand the world around them through their own mental and physical interaction with it?
- involve the use of simple, safe and familiar equipment and materials?
- involve resources and strategies accessible to all teachers?
- give opportunities to work cooperatively and to communicate scientific ideas to others?
- contribute to a broad and balanced science curriculum, bearing in mind other experiences already selected?

These are all important criteria and present an exciting scenario for development. We are certain that all the approaches detailed in our case studies meet these criteria with something to spare. However, once over that hurdle there is still the added problem of actually how to choose between teaching approaches. This gives rise to a subsidiary question: 'Shotgun blast' or 'tuner-receiver' policy? By 'shotgun' approach we mean the use of an array of teaching methods somewhat indiscriminately – relying on the fact that some of the 'shot' will hit some of the targets some of the time. A 'tuned' approach means that the teaching method is carefully matched to the content of the lesson.

Put another way, our question becomes: Is it possible to stipulate that specific topics in the curriculum are most effectively taught through particular teaching strategies? If, for example, we want to teach about osmosis, is there one classroom approach that surpasses all others in its effectiveness? The answer to the question is important because it helps to indicate some of the roles the teacher might take. For instance, if the answer is 'no' then it might indicate the shotgun approach to teaching. That is, any class has within it some 20 or so individuals, all of whom come to lessons with different needs, predilections, preferred modes of learning, and so on. These are not constant and can change even as the session progresses. The best solution might be to pepper the group with a wide range of techniques on the basis that the scatter will have appeal to someone somewhere along the line. Arguably this is the approach taken by most teachers and most courses. Why are some topics turned into practical lessons rather than demonstrations? Why are some set as a homework project rather than group discussion? In published courses there is seldom a discussion of the merits of each approach: 'The best way to achieve understanding in this topic is to . . .'

There again, if the answer is 'yes', and methods can be 'fine-tuned' to meet the needs of certain topic areas, then this would make parts of school life much easier. Each science topic could then come hand-in-hand with optimum technique, and much of the uncertainty about success and achievement would be minimized. And would that be so difficult to achieve? Presumably time-honoured methods work simply because they have been seen to be successful over time with many different groups.

The answer to our question is probably both no, and yes. 'No', because of the human factor. Any group will have among its number a range of approaches to learning, and there has been no strategy invented yet that has guaranteed universal appeal – no matter what the topic area. Every teacher, too, comes with certain strengths and predispositions – there has been no 'teacher-proof' method invented either. 'Yes', because some approaches are clearly superior in tackling some issues than others. Computer simulations, for example, are very good for the exploration of entirely frictionless arenas, and the effects of forces on idealized bodies that lie within them. Something that would prove more taxing for mime or role-play to develop.

In some senses, the search for 'ideal methods' is doomed to be fruit-less. Pupils, schools and teachers are all so different that attempts to 'programme' learning for everyone can be ruled out simply on pragmatic grounds. Philosophically, too, it would contradict the essence of what we, and others, seek in explorations of active learning. Some (all?) of the responsibility for learning rests with the learner and it is they who should influence, if not direct, the shape of the lesson. That said, the content of school science is not infinitely large. In recent years it has been con-strained by a number of interest groups (see Watts and Michell, 1987) and will soon be fairly well detailed through a combination of National Criteria, National Curriculae and GCSE syllabuses. The range of possible teaching strategies is not infinite: schools, classrooms and local communi-ties impose their own sorts of constraints. Moreover, journals like the *School Science Review* are littered with articles by enthusiasts who describe their own tricks-of-the-trade as they design apparatus, experiments or demonstrations to 'get over' various concepts – equipment catalogues are full of their commercial variants.

This has been a trend over many years where, for instance, teacher training 'methods'(!) courses highlight particular ways of working in classrooms. Curriculum materials like the *Nuffield Teacher's Guides*, for example, came very close to operating a 'horses for courses' approach. If you adopted a Nuffield course, you invariably bought the apparatus labelled 'Nuffield' and undertook the activities suggested in the text for the topics to be covered.

Our response to all this is as follows: it is an exclusive policy, not an inclusive one. We would not argue for one-topic-one-method: for any one specific topic to be necessarily tied to one strategy seems far too constraining. We would, however, argue that some areas of work are definitely *unsuited* to particular methods. Teaching sex education in the context of family relationships, for example, is probably not best done by computer programs; computing the acceleration of a trolley by ticker-tape timer is probably not suited to mime techniques. Our message is that teaching approaches should be compatible with the aims of the lesson, or the aims of the course, rather than the subject matter. In this sense we return full circle to active learning and the learning environment. If teachers share the ideals of active learning then some strategies – like the prolonged dictation of notes, continuous chalk-and-talk, recipe-science worksheets – are immediately ruled out, and the search for practical alternatives is on.

HOW CAN WE COME TO TERMS WITH ACTIVE LEARNING?

In Chapter 1 we listed the characteristics of active learners as being students who:

1. Initiate their own activities and take responsibility for their own learning.
2. Make decisions and solve problems.
3. Transfer skills and learning from one context to other different contexts.
4. Organize themselves and organize others.
5. Display their understanding and competence in a number of different ways.
6. Engage in self- and peer-evaluation.
7. Feel good about themselves as learners.

We also suggested that active learning requires other ingredients to make it work:

1. A non-threatening learning environment.
2. Pupil involvement in the organization of the learning process.
3. Opportunities for learners to take decisions about the content of their own learning.
4. Direct skill teaching.
5. Continuous assessment and evaluation.
6. Relevance and vocationalism.

We are not suggesting that this list is exhaustive, i.e. that to simply recreate these would ensure a room full of active learners. The items on the list are necessary but not sufficient. The missing ingredients, however, are not straightforward. One important one, certainly, is commitment. There are several strands to this: commitment by the classroom teacher to try something new, to try to approach some – if not all – of the other characteristics of active learning. It is a commitment to reflect on ways of working in the classroom, and to evaluate past and present performance.

George Walker (1976) gave voice to a simple and oft experienced law of education which states that: 'all educational change takes place in the direction which makes life more difficult for the teacher'. In many ways this book can be read as yet another confirmation of that rule. At the same time it is an attempt to make things easier, by providing ideas and information, making learning more active and enjoyable, and teaching more fascinating and fun. In Chapter 1 we noted some reservations that might be raised against adopting the kinds of approaches we have included in these case studies. The first is time – some practitioners argue that there is insufficient time to cope with even the basic requirements in schools without embracing 'fancy stuff'. There's no doubt that many of the approaches listed here do take more time. They take more time in preparation, in particular for those who have not used such strategies before. They also take more 'real time', i.e. time in the classroom, even if you have used similar techniques before. In our view, this is time well spent. Both teacher and taught have much to gain from exploring new ways of working. Examinations and assessment, too, put considerable constraints on what happens in lessons. There is a clear move towards a framework of regular assessment of youngsters' attainment throughout their school life (DES, 1988). This, however, raises two points. First, assessment these days demands more and more of both the student and the teacher. Different techniques require different kinds of responses from candidates. Perhaps time spent early on will help learners understand, and own that understanding, which will in turn be repaid in increased competence when youngsters come to perform examination tasks. This is not a pious hope. Examination course-work elements demand a variety of ways of working, from project work to oral skills, and attempting skills training in year five seems somewhat late in the day. Secondly, while assessment throughout the school years is becoming a fact of life, work in the early years of schooling can still benefit from variety and excitement in classroom activities:

> pupils start their courses with high expectations: they enjoy working in specialist rooms and being involved in practical work. They try very hard to please, especially in the early years. . . . However the pupils seen were more often bewildered than enlightened by what they did and saw. They were generally not challenged sufficiently by the work. (HMI, 1987)

But what about the conservatism of pupils? There must indeed be numerous examples where youngsters have had changes foisted upon them and reacted with distaste and resentment. Our evidence points somewhat in the other direction. None of our contributors report chaos and revolution; this is hardly surprising, since they are reporting on methods that have worked for them. However, we take this to imply that good practice and innovation can go hand in hand. Where pupils have experienced exciting and challenging approaches to the otherwise predictable everydayness of exposition and recipe science, there has been nothing but engagement and enthusiasm. There is an argument that would suggest curriculum development means curriculum development for pupils too (Ruddock, 1983). We certainly see value in sharing with pupils the objectives and the methods for classroom activities, even if that simply entails talking through with them the programme over the next few weeks.

If teachers interpret George Walker's (1976) notion of increased difficulty as 'more work', then the suggestions contained in the case studies will indeed make life more difficult. If, on the other hand, life is made easier by pupil involvement in the task, their ownership of problems, increased skill at group work, greater self-reliance in project work, or fuller autonomy in the use of resources, then there is much to be gained from the advice herein.

Changes for teachers are difficult too. As Donna Brandes and Paul Ginnis (1986) say:

> The movement from established well-known ground to explore new teaching strategies is a tough challenge for any of us. It requires more and more courage as we get older because we have more to lose in terms of personal status, dignity, self-esteem – especially if we make mistakes along the way.

HOW CAN WE GAUGE THE SUCCESS OF A PARTICULAR TEACHING METHOD?

This is not an easy question to answer because what works for one group or one teacher may not work for another. As good professionals we must rely on our own considerations as 'reflective practitioners', an issue developed by many writers, in particular Donald Schon (1983). We feel it will be almost inevitable that where teachers try a new approach they will compare it to their usual ways of working. It is only by requiring teachers to be reflective of their purpose, principles and practice, we believe, that real progress and change can be made. Obviously, for anyone involved in changing their practice there is one important question: how do you know if the chosen method is working? Part of the answer lies in the aims of the work. If the aims are being achieved then clearly something is going right!

However, this presupposes that the aims are broad ones, not limited. Having, for example, an aim that 'pupils should enjoy their science' is all well and good, but if this were the only aim of a lesson then other aspects of pupils' education would be neglected. Even having an appropriate set of aims is not sufficient. There still needs to be a measure of their achievement. What methods can teachers employ to explore the degree of success of a particular way of working? The first and most obvious (yet often most neglected) step is to ask the pupils. We asked our contributors to provide information about what the pupils achieved from the experiences described. They made points such as:

- all pupils found areas in which they could succeed and in which they could appreciate the work of others;
- most pupils discovered that they were much more capable than they realized;
- all pupils learnt new skills. For example, they learnt communication skills, planning and pacing, working to deadlines, competitive skill and working with the minimum of supervision (case study 2);
- practical skills had clearly been developed, but the pupils had been involved in a lot more. Such things as the ability to work with others, the demands of team work and taking responsibility for the direction and planning of work (case study 3).

There are many ways to collect evidence from pupils. An end-of-unit or termly test can tell us many things about their achievements but, perhaps, little about which teaching approaches were useful, or which one they considered to be most valuable. Techniques such as questionnaires, individual learning logs, or group-written class logs can document their progress and impressions. These can all give youngsters the opportunity to influence directly the shape of lessons – they are shrewd judges of effective teaching approaches. Pupils are, of course, the only people to experience the total school curriculum 'in action'. They frequently see similar topics being taught in a variety of ways, in a range of classrooms and in different subject areas. They know, in their own terms, 'what works and what doesn't', and yet usually have no legitimate mechanism for sharing that evaluative knowledge with teachers. As professional evaluators of classroom experiences, HMI make regular use of the pupils' views of the curriculum by 'shadowing' individuals during school inspections. It is not an opportunity commonly open to teachers. For our purposes, however, we could do a lot worse than listen to what pupils have to say – for them lessons represent much of the everydayness of school life.

WHAT OF TEACHING FOR EQUAL OPPORTUNITIES?

As we note above, one important criterion to be met is that learning experiences should: 'appeal to both boys and girls and to those of all cultural backgrounds' (DES, 1988). There has been growing concern for the place of girls in science education for many years. Projects such as GIST (Girls into Science and Technology – Whyte, 1986) have dealt specifically with the problems faced by girls in science and have indicated that teachers must take gender issues very seriously. For instance, not only must all teachers reconsider their attitudes to girls in science classrooms, they must also devise new strategies for teaching science if they are to ensure that girls develop and maintain an interest in the subject. Some of our case studies make special mention of girls, and we see those studies that emphasize confidence-building and raising learners' self-esteem to be ones which will be of particular benefit. In our view, active learning approaches can do a great deal to enable girls to provide ideas and solutions to scientific problems that draw upon their particular experiences. However, this is but a first step. As we have argued elsewhere (Bentley and Watts, 1986), simply to build confidence is not enough. As Alison Kelly (1987) points out, confidence-building was an approach which, in the early days of research into girls and science, was seen to be most reasonable. Girls opted away from science, it was said, because:

> there must be something wrong with their perceptions of science, the world or of themselves. The corollary of this was that intervention strategies were designed to boost girls' confidence and correct their misconceptions of science.

Alison Kelly herself has made a significant change in her thinking. In 1981, for example, she construed girls' disenchantment with science as:

> girls see science as a difficult subject, and have less confidence than boys in their own abilities; they see science as masculine, which conflicts with their own developing sense of femininity; and they see science as impersonal, whereas their socialisation has primarily been towards concern for people.

She admits now (Kelly, 1987), however, she would construe the issue differently:

> I would put more emphasis on the role of schools and teachers in dissuading girls from science, and less on girls' internal states . . . I now think it is necessary to change science.

This book has not featured approaches to girls and science education as a specific chapter. With Alison Kelly, we believe that treating girls as though *they* were the problem, and designing 'girl-friendly' approaches to science, is not the most successful way to ensure women have equal opportunities to impose their ideas on existing scientific frameworks. Instead, we have provided a menu that should give girls an active part in determining their own science education (and their own science?). Alison Kelly (1987) states that 'planning the route is as important as knowing the destination'. We hope our chapters provide a map from which that route plan can take shape.

MULTI-ETHNIC AND ANTI-RACIST SCHOOL SCIENCE

A strong criticism of science teaching is that it often portrays an image of science that is acultural. That is, science, by its very nature, stands outside cultural influences. We are not of that opinion. As we have suggested elsewhere (e.g. Nott and Watts, 1987; Bentley, 1987) science, and particularly science education, is a strongly cultural activity – it operates, and can only gain credence, through what might be called its 'social licence'.

One or two of the case studies have tackled science in a multicultural context. None of them, though, have specifically directed themselves to this as an objective. Like many other concerns within the studies, anti-racist school science can fill a book on its own (see Gill and Levidow, 1987). There are, however, some general points to be drawn from many of the cases which will help classes to work in a multicultural context.

For instance, where teachers generate conditions for active learning, pupils will gain by having particular kinds of expertise to contribute and, therefore, feel that they and their cultural background are valued. Teachers themselves will gain by having access, through their pupils, to a wide variety of examples and life experiences. Youngsters' experiences, and solutions to problems, always need to be received with thoughtful consideration and positive encouragement. In this case, pupils with differing cultural experiences have the option of drawing on a wider range of possible solutions. The opportunity to explore these solutions openly with peers, and to find that they are given proper value and weight by the teacher, will boost confidence and help to develop full competence in using their science.

We must say, though, that generating genuinely multicultural and anti-racist science education needs much more than this. As Christine Ditchfield (1987) states:

> If we fail to represent cultural variety in the resources used in science lessons, then aren't we denying the existence of ethnic minorities in our society and excluding their experiences? . . . If we fail to include the contribution of many diverse cultures to the development of science, we give the impression that no contribution was made. . . . By broadening the base of science education to show:

- the varied use of science across the world;
- the evolution of scientific ideas and scientific methods through-out history;
- the contribution of many cultures to the international activity of science;
- the limitations of scientific explanations;

we are provided with the opportunity to widen the educational opportunity of all youngsters.

Christine Ditchfield goes on to make many further suggestions for how teachers might provide a fully developed multi-ethnic approach to education. Some of her statements about classroom activities are pertinent to parts of this book. Some examples of anti-racist science teaching – and the debate about anti-racist versus multicultural education – can be found in Gill and Levidow (1987).

WHERE TO NEXT?

Here we want to explore one of the many aspects we have mentioned as important to active learning: a positive, non-threatening learning environment. We see this as a basic requirement for all of us who look for a fundamental change in the way we work. Our description of active learning is a distillation of a range of comments from a variety of sources. An interesting source, for example, has been Brandes and Ginnis' (1986) guide to student-centred learning. The guide derives from the authors' work with TVEI groups in schools in Birmingham, describes the changes that can be made and illustrates some of the techniques that can be used.

They list many ways of starting out on a participatory approach with students. Several have the common thread of being about the classroom atmosphere, the relationships between youngsters in the group. They characterize this by students:

- sitting together instead of in rows;
- speaking openly and honestly with each other;
- listening respectfully;
- sharing unconditional positive regard;
- engaging in joint decision making;
- being free to participate or not.

Donna Brandes and Paul Ginnis (1986) say:

> If all these conditions are operating, no matter what activities are being used, the environment will now be ready for the students to take their rightful position at the centre of their own learning experience.

That is, classroom practitioners who want to encourage active learning fully, need to produce an environment and atmosphere which are consistent with their aims. Careful attention to detail and concern for 'pupil-friendly' activities or curricular materials is not enough by itself, the relationships within the group and the learning environment must be right.

It might be possible, for example, for some of the activities described in earlier chapters to be taken and used in thoughtful ways, but still retain many of the characteristics of transmission teaching and passive learning. It is salutary, for example, to note Olson's (1981) discussion of teachers involved with the Schools Council Integrated Science Project (SCISP) materials. SCISP places a strong emphasis on 'low-influence' situations: the development of autonomous learning in the student so that the teacher's role is more managerial than directive. The teachers reported in Olson's study were enthusiastic about the project materials and the student activities, but not about relinquishing their usual 'control' over events. They were unenthusiastic about such ways of working and did not adopt them in the classroom. While they liked the idea of the project materials, they disliked their 'low-influence' roles as alien, making them – as teachers – feel uncertain and unclear.

In this way, as in many other ways, classroom teachers are the gatekeepers to modes of learning – active or otherwise. As Postman and Weingartner (1971) say:

> There can be no significant innovation in education that does not have at its centre the attitudes of teachers and it is an illusion to think otherwise. The beliefs, feelings and assumptions of teachers are the air of a learning environment; they determine the quality of the life within it.

In our own view the design of the learning environment should be firmly in the hands of both youngsters and their teachers. It is upon this cornerstone that we would argue that classroom activities should be developed. This personalization and negotiation of the learning environment is compatible with views we have worried about elsewhere (e.g. Watts and Bentley, 1984, 1986, 1987; Bentley and Watts 1986). The Secondary Science Curriculum Review (1984) expressed similar concerns:

> The school should encourage young people to become involved in decisions concerning their own learning needs, giving them direct experience of the responsibility which is necessary for their development as self-reliant and self-directing adults. This means the provision of . . . adequate time for students and teachers to talk and share ideas [and] an operational plan for individuals' learning experiences which . . . takes account of the students' earlier experiences.

For us, this describes a view of learning in which the learners are firmly in control of their own learning. Teachers are the facilitators and guides of pupil experience but only the youngsters themselves can construct their personal knowing. This must be based upon their personal experiences, their prior understanding of ideas, their feelings and their preferred mode of learning – plus a classroom atmosphere and activities that are compatible with all this. As Carl Rogers (1969) says:

> The goal of education is the facilitation of change and learning. Teachers should free curiosity and allow learners to go charging off in a new direction dictated by their own interests. In the achievement of this goal it is not teaching skills or lecture or curricular resources which are of greatest importance, but the existence of a personal relationship between facilitators and learners. In a classroom climate characterised by realness, trust, and empathy, learning of a superior quality is occurring.

A LAST PIECE OF ADVICE

There are a lot of good ideas in our case studies. There are some to try at once by yourself in the classroom; some, though, need considerable planning and help from colleagues and pupils. Before jumping in with both feet, think about what needs to be done and why. It is easy to damage relationships with other members of staff by thoughtlessness, and for them to then discredit your efforts at change. Our experience, and the advice that seems endemic in many of the case studies, is to find allies in the system. They may be friends within the school or department, colleagues in other schools or institutions, authority advisers or inspectors. Other people need encouragement too, you will not be the only one, and they will draw strength from your ideas and support. Our advice, then, is find a friend (or friends) to talk and work with. Trying something radically new can mean needing support at crucial times. Active learning can mean active learning for teachers, and an ally is a useful practical alternative to working in isolation. We leave the last word with, this time, Wes Richards' class:

> 'But there might be alcohol in the vinegar. I mean it's called wine vinegar.' Joseph was speculative.
> 'Yes, but they'd have told us about that when we went round the vinegar plant. They never said it had alcohol in it, just that it had a different flavour if it was made from wine rather than from malt.' Christine stated this firmly.
> 'But we're not certain are we?' Joseph responded, 'Scientists don't just take things for granted, they test out the ideas. What we should

do really, is to test all the wines for both alcohol and acetic acid, and then all the vinegars for acetic acid and alcohol.'

'I can't remember what acetic acid is,' said Kate, 'I'm confused – I've missed two weeks, remember. Last time we'd just finished doing the alcohol series, and Mr Richards set that homework on uses of alcohol.'

'OK' said Po Chui, 'in the next lesson he showed us a computer program for looking at different alcohols and what they're used for, and we did some work on the effects of alcohol on the body – you know, reading some stuff about how alcohol affects people, and the sort of problems it creates. Some had to do plays or mimes at the end of the lesson on the "big questions" or problems of using alcohol. You can see what we did anyway, you can look it up in the Class Log. It was my turn to write what we did and put some comments on the lesson.' Po Chui was impatient.

'Mick's group did a great poem on having a hangover. John asked what caused hangovers and someone said that you got worse hangovers if you mixed your drinks or drank red wine. There was something in the red wine that made it worse. So that started everyone arguing and Mr Richards said we'd better see if we could find out if there was anything in the statement. He got us to think up ideas of what it might be that was causing the difference – if there was one – and investigate it. Dave decided that it might be vinegar that was causing the problem. You know maybe a bit of the wine was bad . . .'

In another corner . . .

'OK, so now we've tested all these red wines for methyl and ethyl alcohol are we any the wiser?' Anna asked, 'I mean have we got a firm conclusion?'

'Not really . . . I mean yes we have, but we need to know which ones cause hangovers – or then again perhaps it is a colour problem – you know something to do with the redness of the wine – we'd need to do a chromatogram.' Dennis sounded depressed.

'Perhaps we could predict which ones would cause hangovers and get Mr Richards to try them for us?' They all laughed at the thought of that. 'You know what he'll say – it won't be a fair test, 'cos he can't take his drink anyway!'

'Perhaps we could try them ourselves and design criteria for our Group Skills Assessment – on ability to cope with hangovers?' Denley laughed.

In another part of the room . . .

'Look, perhaps we could mime the results,' Paula suggested.

'But is a mime going to tell people about our results? I mean how are we going to show how many people got hangovers from drinking only red wine, how many of them had migraines, how many never drank at all . . . mime isn't the best way. Perhaps we should write a short story . . . Miss Wilson from English would help us . . .'

Wes Richards smiled as he listened to the conversation. He looked around the room at other groups similarly employed in measuring, arguing, testing out their ideas and theories. It made science teaching so much more interesting when the pupils were enthusiastic like this. It had been a good idea, too, to teach the alcohol part of the syllabus at the same time as the English department's work on advertising. There had been some excellent opportunities for team teaching. He wished the visit could have been to a whisky distillery rather than a pickling factory. Now where was that article about alcohol-driven cars in Brazil . . .

BIBLIOGRAPHY

Bentley, D. (1987). What do you think of it so far? Some youngsters' views about multicultural science. In *Better Science: Working for a Multicultural Society* (Ed. C. Ditchfield). London: Heinemann and ASE.

Bentley, D. and Watts, D. M. (1986). Courting the positive virtues: A case for feminist science. *European Journal of Science Education* **8** (2), 121–34.

Brandes, D. and Ginnis, P. (1986). *A Guide to Student-centred Learning*. Oxford: Basil Blackwell.

Department of Education and Science (1987). *National Curriculum Science Working Group. Interim Report*. London: DES.

Department of Education and Science (1988). *National Curriculum Task Group on Assessment and Testing. A Report*. London: DES.

Ditchfield, C. (Ed.) (1987). *Better Science: Working for a Multicultural Society*. London: Heinemann and ASE.

Gill, D. and Levidow, L. (1987). *Anti-racist Science Teaching*. London: Free Association.

Her Majesty's Inspectorate (1987). *Report by HM Inspectors on a Survey of Science in Years 1–3 of some Secondary Schools in Greenwich*. London: DES.

Kelly, A. (ed.) (1981). *The Missing Half: Girls and Science Education*. Manchester: Manchester University Press.

Kelly, A. (1987). *Science for Girls*. Milton Keynes: Open University Press.

Nott, M. and Watts, D. M. (1987). Towards a multicultural and anti-racist science policy. *Education in Science* **121** (January), 37–8.

Olson, J. (1981). Teacher influence in the classroom: A context for understanding curriculum translation. *Instructional Science* **10**, 259–75.

Postman, N. and Weingartner, C. (1971). *Teaching as a Subversive Activity*. Harmondsworth: Penguin.

Rogers, C. (1969). *Freedom to Learn*. Columbus, Ohio: Charles C. Merrill.

Ruddock, J. (1983). Inservice courses for pupils as a basis for implementing curriculum change. *British Journal of Inservice Education* **10** (1), 32–42.

Schon, D. (1983). *The Reflective Practitioner*. London: Temple Smith.

Secondary Science Curriculum Review (1984). *Towards the Specification of a Minimum Entitlement: Brenda and Friends*. London: SSCR.

Walker, G. (1976). Individualised learning – experiment or expedient? In *Towards Independent Learning in Science* (Ed. E. Green). St Albans: Hart-Davies Educational.

Watts, D. M. and Bentley, D. (1984). The personal parameters of cognition: two aims in science education. *Oxford Review of Education* **10**, 309–17.

Watts, D. M. and Bentley, D. (1986). Methodological congruity in principle and practice: A dilemma in science education. *Journal of Curriculum Studies* **18** (2), 167–75.

Watts, D. M. and Bentley, D. (1987). Constructivism in the classroom: enabling conceptual change by words and deeds. *British Educational Research Journal* **13** (2), 121–35.

Watts, D. M. and Michell, M. (1987). *Better Science: Choosing Content*. London: Heinemann and ASE.

Whyte, J. (1986). *Girls into Science and Technology*. London: Routledge and Kegan Paul.

INDEX

ABAL, 105, 106
active learners, 14, 101, 178
Adams, D., 84
Ashmore, A. D., 82
assessment, 97, 99
  credit accumulation, 9
  graded tests, 13
  practical, 28, 30
  profiling, 9, 13, 119
  records of achievement, 13
  self, 107, 109
  tests, 13
Assessment of Performance Unit
    (APU), 11, 45, 80, 95, 125
Association of Science Education
    (ASE), 7, 44
attainment targets, 13, 184
attitudes, 21, 24, 106, 108, 144, 148,
    164, 191, 194
autonomy, 23, 38, 138, 157, 189

balanced science, 7, 91
Baldwin, J., 158
Barnes, D., 4, 13
Beatty, A., 21, 22
Bell, B., 11
Bentley, D., 13, 163, 165, 179, 191, 194
Billing, P., 13
Black, P., 82
Bolton, G., 153
brainstorming, 36, 60, 76, 168
Brandes, D., 123, 189, 193

Bratt, P., 165
Bronowski, L., 170
Bubel, B., 71
Button, L., 43

Carré, C., 4–6, 45, 74, 79
Casey, R. J., 82
Cawthorn, E. R., 6
Chaundy, D. C. F., 140
classroom atmosphere, 119
Cockcroft Report, 6
competence, 138
computers in science, 186, 187
conceptual change, 11, 12, 54, 75, 95
consolidation, 8, 108
contracts, 118
course-work, 13, 23, 45, 169, 188
CPVE, 80, 82, 83, 176
criteria, 4, 15, 29, 33, 34, 67–9, 75, 77,
    95, 98, 139, 165, 177, 181, 184
cross curricular work, 27, 75, 85, 87,
    95, 111, 132–5, 144, 147
Cussans, B., 91

decision making, 14, 15, 24, 35, 37, 39,
    60, 74, 105, 123, 133–6, 138, 154,
    178, 187
Department of Education and Science
    (DES), 80, 188, 184, 191
discussion, 8, 32, 36, 42, 43, 46, 48–50,
    56, 58, 61, 64, 74–8, 93, 94, 124,
    129, 130, 137–9, 144, 168

differentiation, 24, 136
Diploma in the Practice of Science
    Education (DPSE), 105
display, 15, 16, 129
Ditchfield, C., 192, 193
Dodgson, E., 144
Doherty, M., 110
Driver, R., 11
Dunn, G., 163

Easley, J., 11
educational drama, 17, 142–61
Eggleston, J., 5, 10
Ellington, K., 11
Elliott, H. G., 82
emotions, 18, 112, 144, 148
energy, 71, 72, 85
engineering, 111
Engineering Council, 80
Erickson, G. L., 11
evaluation, 27, 53, 109
    peer, 15, 24, 27, 29, 74
    self, 15, 74, 105, 106
    task, 19
Evans, T., 144
examinations
    'A' level, 105, 106
    GCSE, 6, 13, 23, 24, 27, 45, 113, 165,
        173, 176
experimental design, 22–4

feminism, 112, 116
fieldwork, 27
Fox, D., 5
Frazer, M. J., 82
Freyberg, P., 11, 12
Further Education Unit (FEU), 82, 83

Galton, M., 5, 10
Gilbert, J. K., 11, 47
Gill, D., 192
Ginnis, P., 189, 193
girls and science, 28, 34, 38, 95, 96,
    106, 112–16, 191–2
Green, E., 105
group work, 9, 24, 32, 48, 54, 56, 61,
    71, 74, 76–8, 85, 91–4, 96, 97,

100, 104, 113, 127, 128, 134, 137,
    150, 154
Guesne, E., 11

Hannon, V., 9
Harrison, G., 82
Harvard Physics Course, 64
health education, 133–8
Hinton, K., 82
HMI, 7, 22, 188, 190
Hofstadter, D. R., 64
homework, 106, 107, 110, 151
Howlett, A., 13

ILEA, 105, 106
Ingle, R., 8
investigations, 7, 8, 11, 12, 22, 24, 37,
    52, 84, 87, 91, 99, 114, 117

Jenkins, D. A., 140
Jennings, A., 8
Johnson, K., 63
Jones, K., 124, 139

Kahney, H., 81
Kelly, A., 191, 192
knowledge, 4–6, 12, 61, 83, 97, 109,
    117, 148, 156

LAMP project, 10
language
    bilingual pupils, 56, 59, 77, 96, 170
    in science, 42, 45, 54, 154, 157
learning
    active, 3, 14–16, 18, 21–4, 38, 43, 74,
        77, 101, 107, 123, 138, 157, 163–9,
        177, 178, 183–6, 192
    collaborative, 54, 56, 64, 76
    constructivist, 95
    experiential, 9
    independent, 3, 14, 16, 18, 24, 26,
        52, 163, 177, 181
    individualized, 104–6, 108
    needs, 14, 15
    new, 8, 9
    passive, 3, 4, 5, 10
    resource-based, 162
Levidow, L., 192

Lorac, C., 171

Masterton, D., 140
Mathews, B., 82
McClelland, G., 43, 44, 75, 76
Michell, M., 186
Milroy, E., 159
Moore, J. L., 140
motivation, 29, 32, 50, 53, 91, 106–7,
    111, 139, 144, 151, 157, 167, 170,
    173
Munson, P., 80, 81

National Curriculum, 8, 9, 172, 184
negotiation, 9, 57, 76, 105, 111, 114,
    120, 128, 156
Newell, A., 82
Nixon, J., 144, 147
Non-Threatening Learning
    Environment, 125, 134, 187,
    193–4
Nott, M., 192
Nuffield, 10, 21, 25

observations, 22, 29
Olson, J., 194
Open University, 105
Osborne, R., 11, 12
ownership, 12, 14, 32, 33, 51, 82, 189

Phillips, H., 123
Pillner, G., 64
planning, 39, 101
Posner, K., 14, 181
Postman, N., 194
practical work, 8, 21–3
primary
    schools, 60–2, 75
    science, 117–22
problem solving, 3, 7–10, 14, 17, 24,
    31, 32, 80–7, 91, 92, 94–8, 115,
    117, 118, 124, 136–8, 168, 178, 187
progression, 27
project work, 21–4, 28, 38, 40, 177

Ranson, S., 8, 9
relevance, 16, 32, 94–6, 118, 129, 187
responsibility, 12, 14, 18, 23, 27, 34, 38,
    39, 81, 82, 96, 108, 109, 113,
    115–19, 138, 144, 178, 187, 190
Rice, W., 124
Rogers, C., 104, 105, 195
role play, 17, 135, 142–61
Rowell, J. A., 6
Ruddock, J., 189

SATIS, 10
SATRO, 175
Schollar, J., 82
Schon, D., 189
Schools Council Integrated Science
    Project (SCISP), 10, 194
science
    anti-racist, 192–3
    multicultural, 192–3
    process based, 84, 87, 91
    social implications, 18, 88, 91, 123,
        142, 149–51, 159, 163, 165, 169
    and technology, 18, 50, 82, 83, 94,
        118, 123, 128, 149, 163, 165, 169,
        184
Science at Work, 10, 68
Secondary Science Curriculum Review
    (SSCR), 10, 51, 53, 83, 84, 97, 117,
    194
Simon, H. A., 82
SISCON, 10
skills, 21–4, 29, 31, 34, 35, 38, 40, 44–6,
    51–3, 59, 60, 74–6, 83, 85, 87, 91,
    95–7, 158, 171, 187, 189, 190
    organizational, 107
    study, 106
    teaching, 16
    transfer, 14
Snashall, D., 64
Stenhouse, D., 6
Strike, G. J., 14, 181
Suffolk Education Authority, 2

Taylor, C., 142
teacher appraisal, 2, 7, 184
teacher's role, 86, 98, 101, 131, 138,
    162, 185, 194
teaching
    active, 12
    collaborative, 113

teaching – *cont*.
  interpretive, 4, 6
  styles, 7, 28, 72
  team, 119, 147
  transmission, 4–7, 10, 107
technician's role, 26, 30, 36, 40, 92, 95,
  102, 108
Thomas, F. H., 140
Tiberghien, A., 11
Times Educational Supplement, 165
TVEE, 8, 176
TVEI, 8, 173, 193

Understanding British Industry (UBI),
  175

Van Ments, M., 158
vocationalism, 16

Walker, G., 188
Watts, D. M., 11, 47, 163, 165, 179,
  186, 191, 192, 194
Weingartner, C., 194
Weiss, M., 171
Wells, H., 158
West, R. W., 5
Whyte, J., 181
Woolnough, B., 21, 22
work experience, 9, 177
write to learn project, 60

Zylbersztajn, A., 6